EGGS

NATURE'S PERFECT PACKAGE

EGGS

NATURE'S PERFECT PACKAGE

Robert Burton

Photographs by
Jane Burton and Kim Taylor

Facts On File Publications
New York, New York • Oxford, England

First published in the United States of America by Facts On File, Inc.
460 Park Avenue South, New York, New York 10016.

This book was created and produced by
Roxby Natural History Limited
a division of Roxby Press
98 Clapham Common North Side
London SW4 9SG

Editor: Gilly Abrahams
Design: Eric Drewery
Typesetting: Vantage Photosetting Co. Ltd.
Reproduction: F.E. Burman Limited

Library of Congress Cataloging-in-Publication Data
Burton, Robert, 1941–
    Eggs: Nature's Perfect Package.
    Bibliography: P.
    Includes Index.
    1. Eggs.   2. Embryology.   3. Reproduction.   I. Taylor,
Kim.   II. Burton, Jane.   III. Title.
    QL971.B93   1987   591.1′6   86–29182
    ISBN 0–8160–1384–5

Printed in Italy by
New Interlitho S.P.A., Milan

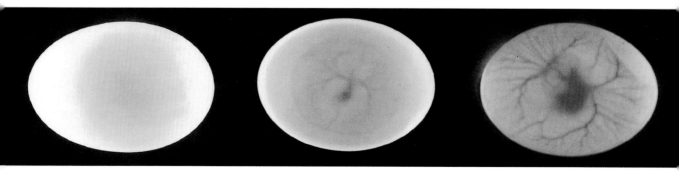

Contents

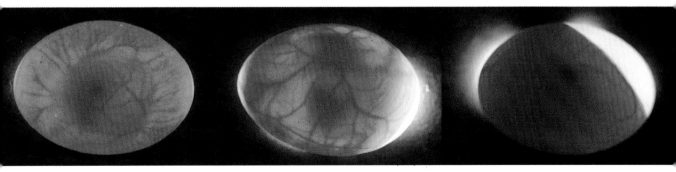

Introduction

The egg is surely one of nature's most remarkable and versatile inventions. It is a compact, self-contained capsule containing everything necessary for the creation of a new life – be it an amoeba, a moth, an ostrich or a human being. For millions of years the egg has carried life from one generation to the next, and many ancient civilizations – Egyptians, Indians and Japanese among them – believed that the world itself had been hatched from an egg made by the Creator. The egg was also a symbol of rebirth and the giving of Easter eggs has been a spring custom in many countries for centuries, linking pagan ideas of the rebirth of nature after the dead of winter with the Christian belief in the Resurrection.

The power of the egg's symbolism is enhanced by its beauty of form and colouring. One hundred years ago, the naturalist T.W. Higginson wrote: 'I think if required on pain of death to name the most perfect thing in the universe, I should risk my fate on a bird's egg.' He should not have stopped at a bird's egg; the tiny sculptured button of a butterfly egg and the translucent pearl of a snail egg have their own beauty. But beyond the mystical and aesthetic appreciation of eggs, there is the enormous biological interest of the strategies employed by so many animals in fertilizing, laying and caring for their eggs.

The variety of eggs makes their definition far from simple. There is no doubt about the egg on the breakfast table; if it had not been stolen from the nest, it might have hatched into a chicken which would have laid more eggs. There is no answer to the old riddle: neither chicken nor egg comes first. They are alternating stages in the never-ending progress of life, except that seen, in a wider sense, eggs came tens of millions of years before chickens appeared on Earth. Birds' eggs are the best known and most often seen, but eggs are found at all levels of the animal kingdom and the story of eggs leads us into some strange corners of natural history. There are octopuses that lay poisonous eggs, insects that lay eggs in other animals' eggs, cuckoos that get other birds to look after their eggs and lizards that lay eggs without sex. Yet throughout the story the egg's fundamental role is unchanging.

The egg is the reproductive unit, produced by the female, which develops into a new individual. It is a single cell, although very different from the other cells that make up the body. No other cell can survive outside the body. In birds the egg grows to an enormous size and the 16-centimetre-long ostrich egg is often said to be the largest cell in existence. It would be nearer the truth to say that the yolk of the ostrich egg is the largest cell. The 'white' and the shell are formed around the egg cell. Yet it is usual to talk of the 'chick developing inside the egg'. Here, egg means eggshell. It is also usual to say that 'birds lay eggs, whereas mammals give birth to babies'. But the processes of embryonic development in birds and mammals – and indeed in the early stages of a frog, a fish, a worm or a sea urchin – are not very different, so why is such a vital distinction made? The answer can be found by going back to the beginning and tracing the start of an animal's existence.

The frontispiece of William Harvey's *de Generatione Animalium* (Concerning the Generation of Animals), published in 1651, shows the hand of Jove holding an egg, or rather an eggshell, out of which have emerged a wide variety of animals: a child, a dolphin, a spider and so on. Worked into the design is the motto *Ex ovo omnia* – everything comes from an egg. Harvey, famous for the discovery of blood circulation, demonstrated that the origin of the developing embryo, and therefore of the complete animal, is to be found in the tiny egg of the mammal. Not having a satisfactory microscope, he was unable to trace the egg right back to its beginnings. This was left to Anton van Leeuwenhoek (1632–1723), the Dutch draper who developed his hobby of making microscopes into an art; his beautifully ground lenses revealed details that no human eye had seen before.

Leeuwenhoek proved by his observations that animals were bred by sexual reproduction and did not arise spontaneously as had been supposed. He showed that grain weevils, for instance, hatched from

After a short period of development, brine shrimp eggs hatch into free-swimming larvae that feed on plankton while they develop into the adult form.

tiny eggs deposited by female weevils rather than arising from wheat grains. Similarly, mussels and other shellfish were not generated out of sand but developed by the same reproductive processes as higher animals. The egg, therefore, provides the essential continuity of life and forms the link between generations. It is virtually universal; only a few animals have renounced sexual reproduction completely and many of these still retain the egg as a means of propagation.

An egg starts as a tiny cell, usually in a specialized organ – the ovary. Like other cells it consists essentially of a blob of protoplasm enclosed in a membrane and containing a nucleus that holds the genetic material – complex molecules of deoxyribonucleic acid or DNA. These DNA molecules form a blueprint made up of units called genes. Each gene controls one small part of the functioning of the cell and together they determine the structure, physiology and often the behaviour of the animal. The nucleus of an egg therefore holds the information needed to build the animal that will develop from it. When the egg is fertilized by a sperm, their genetic materials meet and mingle, so the zygote (the cell that results from the union of the egg and the sperm) contains a mixture of blueprints from each parent.

Our ideas of what constitutes an egg are influenced by the circumstances in which the embryo develops from the zygote. All eggs are supplied with yolk to nourish the early growth of the embryo, but further provision must be made for the main period of development when the body is growing. Many marine, and some freshwater, invertebrate animals have eggs with little yolk, but these transform into larvae soon after fertilization. Larvae are partly formed animals, often totally unlike the adult in appearance, which lead independent lives and can feed themselves when the yolk has run out. Amphibians and some fishes go through this stage of development. Animals which give birth to live young, such as mammals, also provide their eggs with a small amount of yolk but the embryo is nourished in its later stages with food supplied by the mother. A third group of animals provides the egg with enough yolk to sustain the embryo through its development. Such eggs are enclosed in a protective shell and are often laid in a nest. These are the eggs of birds, reptiles, insects, crustaceans, spiders and some kinds of fishes. They are what we normally think of as 'eggs'.

It can be seen therefore that eggs differ in the care that they receive from their parents. In simple life forms they are abandoned to their own devices, while

Left: well-supplied with droplets of yolk, a trout is forming inside its transparent egg. The pigmented eyes are clearly visible.

Above right: the pinhead-sized egg of the brown argus butterfly reveals its perfect sculpturing when magnified.

Right: a Muscovy duckling bursts out of its shell. It had been lying in the wrong position for hatching normally, and would have died in the shell had it not been strong enough to push and kick its way out.

at the most advanced level they are retained in the mother's body where the embryos are nourished, or they are provided with sufficient yolk to complete their development. This last group of eggs may not be merely provisioned. Their parents may fend for them, by guarding and incubating them until they hatch, or the eggs may have their own defences, such as waterproof or camouflaged shells to assist their survival. Whatever the kind of egg, the parents must ensure that sufficient numbers are laid and brought to full development to perpetuate their line.

The whole fascinating story of eggs has not been told before, although it is such an essential part of every animal's existence. The life of an animal does not begin at hatching nor even at laying, but considerably earlier, at conception. From this point a remarkable story emerges of the development of new life from the events of fertilization to the growth of a perfect new animal. The egg has a life of its own – breathing, excreting, growing and even communicating. For an animal to continue its lineage it must produce eggs that are not only adapted to their own environment, but must also be laid in sufficient numbers to survive in a hostile world. And survival is the essence of the story, if a species is to renew itself generation after generation.

CHAPTER 1
Evolution of the egg

Animals have been reproducing for hundreds of millions of years, yet only two basic methods of reproduction have evolved in all this time. There is sexual reproduction, involving the production of an egg, and asexual reproduction, involving splitting or budding. They have continued, unaltered in essentials, almost since animal life began. But while asexual reproduction was the first method to appear and is still common at the lower end of the animal scale, the overwhelming majority of modern animal species reproduce by sexual means and the egg is an essential part of their life cycle. Indeed, we are so used to the coming together of two sexes, not only in our own species but in all familiar animals, that any means of reproduction other than fertilization of the female egg by the male sperm seems rather bizarre. Nevertheless, some animals abandon sex, at least temporarily.

Asexual reproduction is the simpler form of reproduction in genetic terms. When the animal divides into two, it separates its genetic material equally between the two offspring. Sexual reproduction is more complicated because it involves the formation and fusion of egg and sperm – the specialized sex cells or gametes – and the formation of a completely new body. Complex, active animals can only be constructed by going back to the simplicity of the single-celled egg and rebuilding.

The evolution of sex has proved a problem for biologists: why bother to expend energy and body resources producing eggs and sperms for the time-consuming and wasteful process of bringing them together, when asexual reproduction is quicker and simpler? There has to be a positive advantage for sex to have arisen in the first place and it has clearly been worthwhile since nearly all animal species devote time and energy to it.

Evolution is faster with sex than without it because favourable changes spread rapidly through a population as the individuals interbreed. Asexual animals vary only as a result of accidental mutations – changes in the genetic material – which are rare and random events and confined to the descendants of the mutant individual.

In 1859 Charles Darwin published his theory of evolution by natural selection, which showed how the variation imparted by sex allows rapid evolution. The theory states that animals produce an excess of offspring. Those that are best adapted to the environment will survive and breed; those less suited will be 'weeded out'. Because each offspring receives a different mixture of genes from the two parents, they are diversified. So, when there is a change in the environment, those that happen to have the characteristics that are better adapted to the new situation will survive and produce a new line. Without sex and natural selection a change in the environment could wipe out a whole population if they were all identically unsuited to the new environment.

As an example of the advantages of diversity, take the mosquitoes that transmit yellow fever, *Aedes aegypti* and *Aedes africanus*, who breed in temporary puddles in dry country. When attempts were made to breed these mosquitoes for research purposes, it proved difficult to get the eggs to hatch, so a closer investigation was made of their egg-laying habits. Females were given the blood-meal (from the researcher's arm!) that these mosquitoes need to develop their eggs. The eggs within a clutch often took a variable time to hatch; sometimes weeks elapsed between first and last, because each one developed almost to the point of hatching and then stopped. This state of arrested development is called a diapause and it lasts a fixed time, after which a stimulus from the environment causes the egg to hatch. The length of the diapause is inherited and the variation within a clutch of eggs is valuable to the mosquitoes: on the one hand, it is an advantage to hatch as soon as possible so that the adult mosquito can emerge before the pool dries up; on the other hand, if the larvae hatch out when there has been only a minimal amount of rain, the pool may dry up too

Volvox colonies gather at the surface of the water in sunlight with daughter colonies forming inside them. *Volvox* reproduces both sexually and asexually, according to conditions.

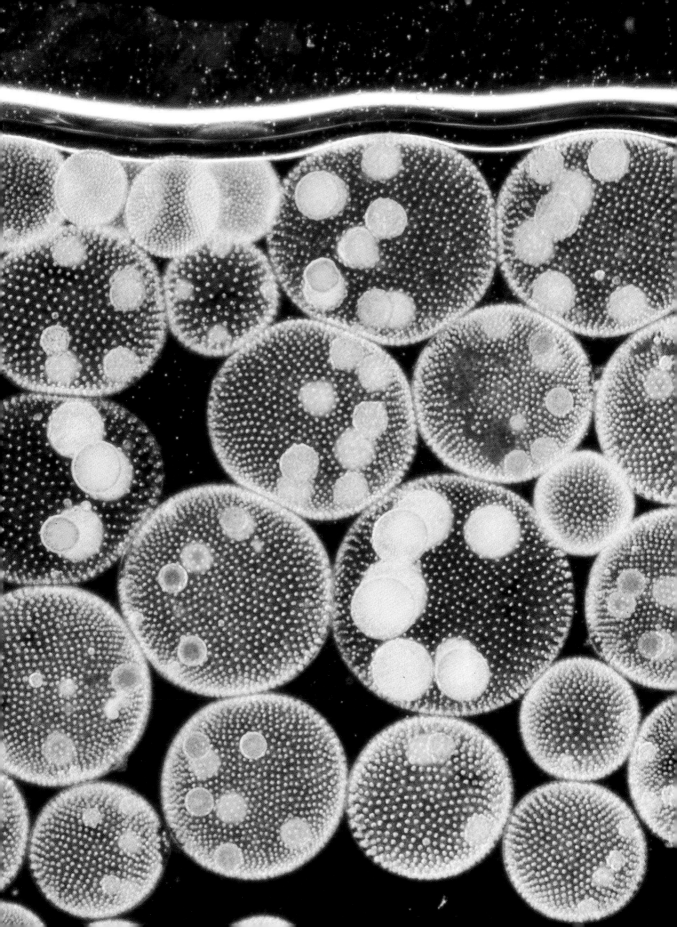

soon. Hedging their bets is therefore a useful survival strategy for the mosquitoes, since a range of hatching times will cover all eventualities.

There is, however, a snag to sex. Because an egg receives genetic material from a male parent, the female has to share her genetic investment. The object of reproduction is to pass on one's characteristics to the next generation, so why help to pass on someone else's? One theory is that, instead of being a source of genetic diversity, sex originally evolved to protect the earliest organisms against potentially disastrous changes brought about by genetic mutations. Without sex there could have been an accumulation of random errors in the genes, but by carrying two copies of genetic material, one from each parent, defects in one copy would be compensated for by the other.

Whatever the reason for the evolution of sex, it is found at all levels of the animal kingdom and the egg in its many forms is its manifestation. To understand the role of the egg in the lives of animals, it is necessary to start with the most basic kind of reproduction, which is simply to split in two. This method, known as binary fission, is used by some species of amoeba, the single-celled animal or protozoan that is often used as an example of the simplest form of animal life. *Amoeba proteus*, for example, starts its reproductive process by becoming spherical, then the nucleus containing the DNA divides to form two nuclei; finally the amoeba splits down the middle, forming two separate but genetically identical amoebae, each with its own nucleus. The whole process takes about an hour.

Some amoebae have evolved a variation of binary fission in which hundreds of new amoebae are formed by repeated division of the original nucleus. By secreting a tough external 'wall' or membrane, each one makes itself into a cyst that can survive if the pond dries up. Then if it is carried by the wind or in mud on a bird's foot to a suitable new site, it can establish a new population. The cysts therefore perform two of the functions of an egg: they maintain the species during adverse conditions and act as a means of dispersal.

More complex forms of animal life are sometimes able to reproduce by a form of fission. Sponges, for instance, will regrow new individuals from pieces

---

Amoeba dividing asexually. Each half will have a nucleus containing identical genes, so there will be no variation between generations except by accidental mutation.

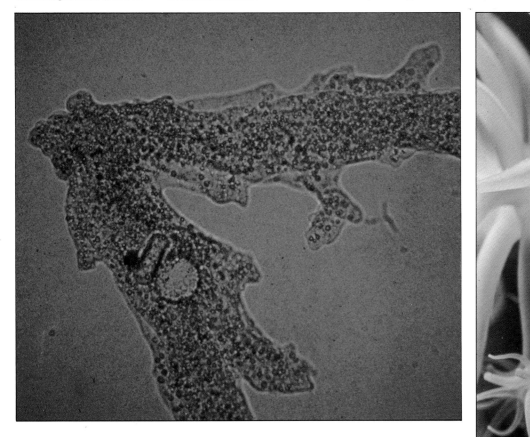

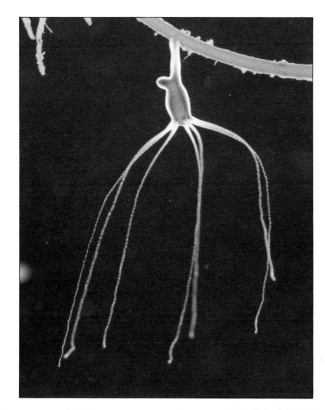

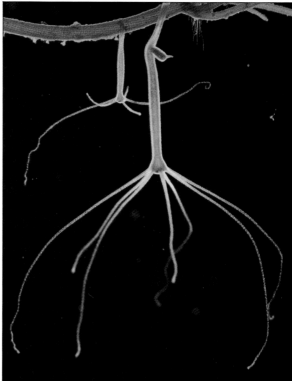

Left: although the beadlet anemone can produce eggs by sexual reproduction, it also reproduces by budding. The buds are formed within the body cavity and escape through the parent's mouth to settle nearby.

Above: *Hydra* reproduces by budding as an alternative to making eggs. At first it is difficult to distinguish a bud from an egg, but the bud soon grows tentacles and eventually separates to become a new individual.

torn off the parent body by a storm. Similarly, sea anemones sometimes have pieces of the pedal disc or foot torn off as they creep over the rocks; the pieces grow into new anemones. In more organized fission, the body of the anemone splits in two, like an amoeba but on a much greater scale, then each new half of the body regenerates the half it has lost.

These simple animals also create replicas of themselves by budding. A sea anemone or *Hydra* – a very similar animal that lives in ponds – grows a new individual out of the side of its body. A small wart-like bud appears on the surface and takes on the form of a new individual, as tentacles and mouth appear. For a while 'mother' and 'daughter' live like Siamese twins with intercommunicating stomachs, then the daughter 'nips off' at the base to become independent.

Some flatworms and bristleworms bud off new individuals or split into two or more pieces which grow into complete individuals, and if you cut an earthworm in two, the separate halves will grow a new head and tail respectively. However, at this level of organization the body is on the point of becoming too complex for asexual reproduction.

Although reproduction by fission is restricted to animals that are not very mobile and have relatively simple bodies, sexual reproduction is by no means the prerogative of advanced animals. A type of sexual reproduction, and the production of eggs, is widespread among the protozoans. At its simplest, two individuals of the slipper-shaped *Paramecium* bump into one another and join together in the process usually known as conjugation. At the point of fusion the cell membranes break down and there is intermingling of the cell contents or cytoplasm of the two individuals. The nucleus of each divides and one half migrates into the other individual, to fuse with the half nucleus which has remained behind. The animals then separate and, soon after, each divides several times to give a clutch of offspring containing the new combination of genetic material.

Other protozoans form gametes not very different from those of higher animals. *Chlamydomonas*, which propels its pear-shaped body through the water by beating two whip-like flagella, is a protozoan that is part animal and part plant; like a plant it contains green chlorophyll to trap the energy from sunlight to manufacture starch from carbon dioxide and water, but it also eats food like an animal when kept in the dark. To reproduce, it divides repeatedly to form several small replicas of itself known as isogametes, meaning 'similar gametes'. They meet and fuse with isogametes from another individual to form a zygote that develops into a new *Chlamydomonas*.

An extension of this behaviour is for the gametes to be of unequal sizes. This is called anisogamy. The larger and less active gametes are formed by a single division of the parent rather than repeated division like the isogametes. The smaller gametes are formed by the same repeated division as the isogametes, and they meet and fuse with the larger gametes from different organisms of the same species.

Anisogamy is the first step in the evolution of the egg; here is the start of the division of labour between the sexes. The larger and less active gamete is, by definition, the egg, and the smaller, more numerous gametes are the sperms. So already in the protozoans the basic events of sexual reproduction and the function of an egg are established. Throughout the rest of the evolutionary tree of the animal kingdom – from these simple, minute organisms to the complexity of birds and mammals – there is no radical departure from this pattern. It merely becomes more elaborate, with asexual reproduction becoming a less viable option.

The first elaboration is seen in the organisms that consist of colonies of *Chlamydomonas*-like cells. One of the simplest of these, *Pandorina*, consists of 16 *Chlamydomonas*-like cells embedded in a ball of jelly with their flagella on the outside. Reproduction is either asexual or sexual. In asexual reproduction, each of the 16 individuals in the colony divides to form a new group of 16 cells. When the ball disintegrates, it releases 16 new colonies, each composed of 16 cells. Sexual reproduction differs in that when the cells divide, the new cells are released separately; they swim around until they meet another cell and fuse. The resulting zygote divides to form a new colony.

Volvox is a larger relative of *Pandorina*. The colony, which is just visible to the naked eye, contains from 500 to 60,000 cells, set in the surface of a hollow sphere and connected to their neighbours by delicate strands of protoplasm. The *Volvox* colony is more than a mere collection of individuals. As it moves through the water, it spins (its name comes from the Latin *volvere*, meaning 'to roll'). This is not a random rolling; it spins on an axis so that it has a 'head end'. The cells at the head end are sensitive to light and guide the colony towards moderate light intensities. These cells do not take part in reproduction, so sexual and asexual division is limited to a few cells at the rear end of the colony. Asexual reproduction consists of one cell dividing repeatedly to form a daughter colony, which swims inside the mother. A brood of daughters eventually collects, straining the mother's body until she tears open and they escape. After several generations of asexual reproduction, the colonies may suddenly start to reproduce sexually.

In some species of *Volvox* there are separate sexes. In others, male and female gametes are formed in the same individual. The eggs are formed by a single cell losing its flagella and enlarging. Sperms are formed by a cell dividing repeatedly, as in asexual reproduction. The sperms then break free of the colony and swim away, either still in a packet or as separate sperms, and penetrate the gelatinous covering of a female colony to fertilize the eggs.

The fertilized egg, or zygote, forms a thick, hard coat which, like the amoeba's cyst, will survive adverse conditions. When the pond or ditch dries up or freezes, the *Volvox* colonies die, but the zygotes survive. With the return of conditions favourable to life, the zygote's 'shell' splits open, cell division starts and the new individual is formed.

There is a range of these colonial creatures, from the 16-celled *Pandorina* to the many-celled *Volvox*. As the number of cells increases, so the proportion that takes part in reproduction decreases. The sequence shows the gradual specialization of tissues to form the vegetative body or 'soma', which plays no part in

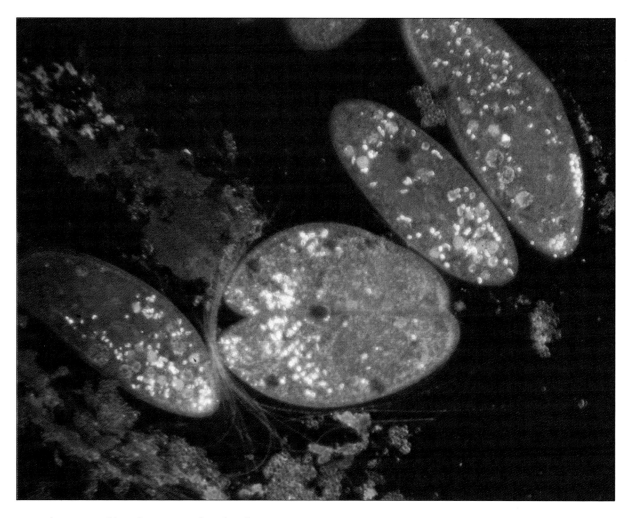

reproduction and has the potential to develop into tissues specializing in digestion, movement, excretion and so on. Only the remaining fraction form a tissue for producing the gametes.

Volvox and its relatives show the way that the mechanism of sexual reproduction could have evolved, but they are themselves an evolutionary dead end. They have, however, evolved far enough to show the essential features of eggs.

The eggs of many different kinds of animals, from birds and reptiles to insects and snails, and including the zygotes of *Volvox* and the spores of *Amoeba*, are provided with a tough, more or less impermeable shell, which protects the tissue of the egg. The shell has been one of the key adaptations for life on land, as will be discussed on later pages, but its main function for many animal types is to withstand temporary adversity.

Animals that have the choice of asexual or sexual reproduction characteristically reproduce by asexual means when conditions are good. Budding and fission allow a rapid increase of numbers so that a species can take full advantage of warm weather when food is plentiful and growth is rapid. Then, as the situation deteriorates, sexual reproduction takes over and eggs are deposited as time capsules to keep the species going through drought or winter.

Hydra, for example, reproduces by budding throughout the summer, but usually switches to sexual reproduction in the autumn. Swellings, similar to the buds that develop asexually, appear on the base of the body. Each lump contains one very large cell which develops into an egg. Similar lumps near the tentacles produce sperms which swim to the eggs. When fertilized, the egg develops into an embryo which secretes a hard, sticky shell around itself, then

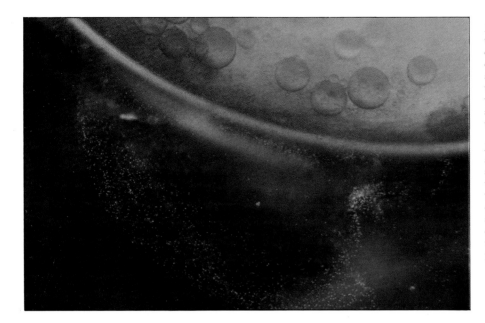

Left: the egg of a rainbow trout, with droplets of yolk clearly visible, is huge compared with the tiny sperms swarming around it. To fertilize the egg, one sperm must force a way through the egg membrane and penetrate to the nucleus in the centre.

Right: *Daphnia* water fleas swarming in fresh water. Some are carrying 'resting eggs' in their abdomens. Produced when the pool becomes overcrowded, these are large and have thick shells which enable them to survive in freezing conditions or if the pond dries up.

drops off the parent *Hydra*. It lies on the bed of the pond or on the leaf of a water plant until eventually it hatches out and grows into a new *Hydra*.

These resistant eggs have a second function because, as we have seen, they can be transported to establish new colonies. Compared with the adult animal, an egg is very small, so it has a better chance of either being carried in the wind or sticking to the body of a larger animal.

The macaw worm, a parasitic fly living in Central America, has a unique method of dispersal. Its maggots feed on the flesh of mammals and the fly uses an intermediary to transport its eggs to a suitable host. It lays an egg on the body of a mosquito, which flies in search of a human or other large mammal and sucks its blood. While the mosquito is occupied, the macaw worm maggot hatches out of the egg in a matter of seconds and burrows into the skin of the mammal.

To sum up, the egg secures the survival of the species through physically difficult times, but for most species its more important function is to be the vehicle for the continuity of the species in the long term. As the repository of genetic material from both parents, the egg carries the variability that gives the opportunity for evolution and adaptation to long-term changes in the species' environment. The mechanism of variability is inherent in the production and fusion of the sperm and the egg. To understand this, it is necessary to know something about how an egg is made and, in particular, the mechanism by which the genetic material, the DNA, is passed on to the next generation.

The making of an egg

The first stage of sexual reproduction is the transformation of cells in the reproductive organs into specialized sex cells, the gametes. In higher animals, eggs start as simple cells in tissue laid down when the mother was an embryo (and the sperms start similarly in the male embryo). Development to a ripe egg, ready for fertilization, can be a lengthy process. All the eggs that will ever be released are laid down before birth, so some eggs of humans, long-lived whales, albatrosses and tortoises may not ripen for around half a century, or even longer in some exceptional cases.

During the early period of development the prospective egg cells, known as oogonia, show little change. Then there is a period of comparatively rapid development both in the structure of the egg and within its nucleus before the egg is laid. Yolk is laid down in the cytoplasm to feed the fertilized egg until it hatches, while changes in the nucleus prepare the egg for fusion with the sperm.

An animal's genes, the units of DNA molecules, are carried in the nucleus of each of its cells. When a cell is about to divide into two new cells, the DNA arranges itself in a bundle of paired threads called chromosomes. This is called the diploid condition. The pairs of chromosomes separate and each chromosome constructs an identical partner. This process of mitosis results in the new cells having genes and chromosomes identical with the original cell. Thus, as an animal's body grows and worn tissues are replaced during its life, the blueprint in the cells remains unchanged.

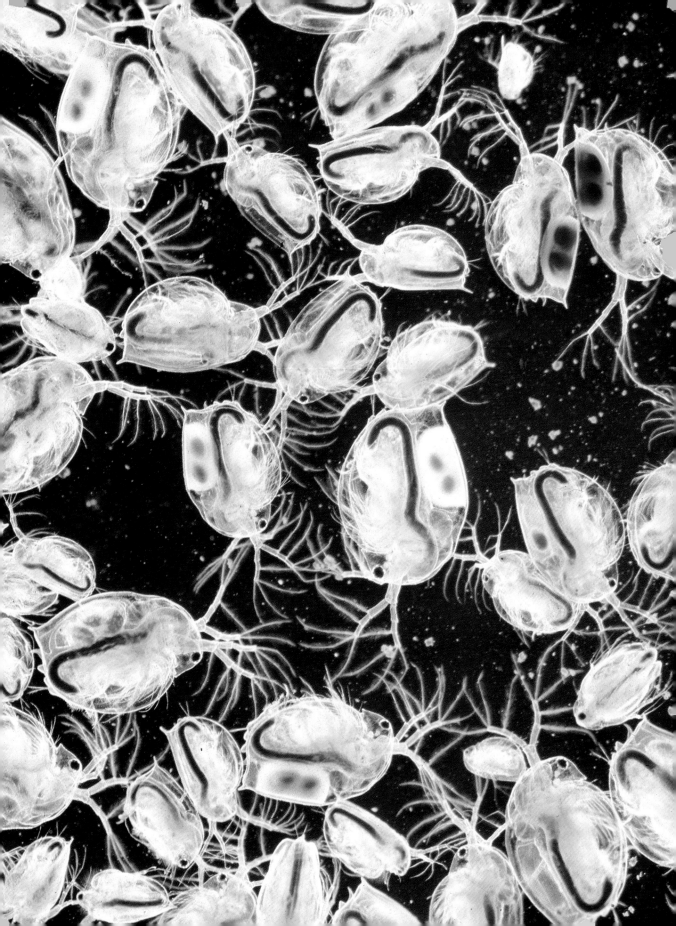

When gametes form, cell division follows a different pattern. The number of chromosomes has to be halved so that after fusion of the egg and sperm they will be restored to the original complement. Cells with half the full chromosome number are termed haploid. In chickens, the chromosome number of body cells is 78. This is halved to 39 in eggs and sperms, and fusion restores it to 78. The crucial cell division by which chromosome numbers are reduced is called meiosis. It starts while the animal is itself an embryo and may not be completed until the sperm penetrates the ripe egg. The sea urchin is a rare exception in which meiosis is completed before fertilization.

The process of meiosis, at its simplest, consists of the chromosomes dividing once while the nucleus divides twice, so that a single diploid cell containing paired sets of chromosomes divides into four haploid cells with single sets of chromosomes. First, the chromosomes duplicate themselves, as in mitosis. The two sets separate into two nuclei, one of which is discarded from the cell as a short-lived polar body which looks like a miniature cell. This division is followed by a second in which the pairs of chromosomes remaining in the nucleus divide into single sets. The nucleus splits in two again, and one half is ejected from the cell as before.

The egg now has only a single set of chromosomes and is ready for fertilization by a sperm in which a similar process of meiosis has taken place. The essential difference is that, instead of losing polar bodies, the dividing male cell retains all the new nuclei and turns into four spermatozoa. At fertilization, the single sets of chromosomes from each gamete unite to form a paired set, so the animal that develops from the fertilized egg will have genes from both parents in every cell of its body.

While meiosis is proceeding, the cytoplasm of the egg is also undergoing changes. The size of the egg is increasing enormously as yolk is laid down. A frog egg swells to such an extent that nearly one-fifth of the female's body becomes taken up with eggs. Changes in the ripening egg are even greater in birds. For example, research has shown that the ovary of a Japanese quail increases from 15 milligrams in winter to 6,500 milligrams in the breeding season, mainly due to the deposition of yolk in the eggs that are being prepared for ovulation. The growing eggs absorb the raw materials for their yolks from nearby cells, like ocean-going ships refuelling from tankers at sea.

Even among the simplest multicellular animals, such as sponges, separation of gametes into sperms and yolk-carrying eggs is established. The cell that is to become a sponge egg attracts other cells that gather closely around it. These are known as nurse cells because the food they absorb is transferred to the egg where it becomes yolk.

In insects, the number of nurse cells attending each developing egg varies from one in earwigs to 48 in queen honeybees. The nurse cells absorb nourishment from the blood and pass it to the egg. In some cases the nurse cells are actually engulfed by the egg. Nurse cells can only supply sufficient food in the early stages so, as the egg grows and moves down the oviduct, it becomes surrounded by a tight sheath of follicle cells. These pass food materials to the egg which are transformed into droplets of yolk. Finally, the whole egg is enclosed in the vitelline membrane.

The process in birds is essentially similar. The follicle cells surrounding the egg secrete oestrogen sex hormones which stimulate the liver to produce yolk from the body's food reserves. So great is the demand for yolk in a breeding bird – 15 grams a day for a chicken – that if a bird does not have enough to eat, egg laying may be delayed or fails altogether. Finally, the yolk material is transported from the liver to the follicle where it is transferred to the egg. When development is complete, the follicle switches to producing another sex hormone, progesterone, which triggers the breakdown of the follicle. The ripe egg is then shed from the ovary and enters the oviduct where it will meet sperms and be fertilized.

Seeds and eggs

Seeds are the plants' equivalent of eggs. They look rather like eggs and, indeed, it is easy to show similarities in their structure and function. Both are made by the female of the species as a result of sexual reproduction and they carry the genes of their parents. They are the 'buffers' between generations, some surviving periods of adverse conditions and carrying the species to new places. There are, however, differences in the internal structure of eggs and seeds which are the result of millions of years of separate development as plants and animals evolved along their different paths.

The plant and animal kingdoms separated about 3,500 million years ago, but because fossils from this era of the Earth's history are extremely rare, we have little idea of what life was like in those days. Biologists believe that animal and plant kingdoms arose from ancestral organisms similar to *Chlamydomonas* that share the characteristics of both animals and plants. Even at this stage of evolution the basic patterns of reproduction had been established and, like animals, plants reproduce both asexually and sexually.

The structure of an unfertilized seed is much more complicated than an egg. It is formed in the ovary or carpel in the centre of the flower where the petals join. Within the carpel there are several ovules and within each ovule there is a single egg cell. In the early stages of plant evolution the male gamete was an independent sperm, but in flowering plants it is now no more than a cell nucleus in the pollen grain. Pollen grains are vehicles for carrying the male gamete to the female. They are produced in enormous quantities and there is as much loss as there is among the sperms of animals. Some are carried by the wind, others are transported by bees, birds or other animals, but only a tiny fraction reach the stigma of a female flower. Those that do complete the journey put out a tube which grows down the stigma and carries the male nucleus into the ovule, where fertilization takes place.

The next stage in the formation of a seed is very different from that of an egg, but the end result is the same. A second nucleus from the pollen grain fuses with two nuclei near the female nucleus in the ovule to make a triploid cell with nuclei containing three sets of chromosomes, instead of the usual two. This cell develops into the endosperm tissue, which is the equivalent of the egg's yolk.

The seed is now fully developed. It consists, like an egg, of an embryo surrounded by food and enclosed by the wall of the ovary which becomes the hard outer coat of the seed, the equivalent of an eggshell. In many plants the seeds remain in the carpel which

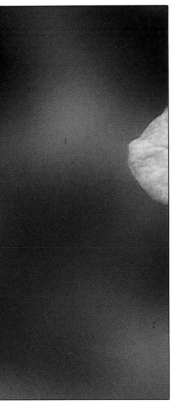

Left: a bee fly hovering in front of a flower shows the pollen grains that it has picked up on its proboscis and legs from the stamens. The grains will now be transferred to the stigma of another flower, thereby pollinating it. Pollination is the process in plants by which the male gamete is carried to the egg to fertilize it. It is therefore the approximate equivalent to mating in animals.

Right: a cloud of pollen bursts from the inconspicuous flower of a stinging nettle. Wind-borne pollen is less certain to reach another flower than that carried by insects, so it has to be produced in larger quantities.

becomes the fruit, an extra protection and food store for the plantlet, like the capsules containing batches of eggs (see page 26).

Also, like the eggs of many invertebrate animals, the seed's tough coat and reserves of food make it a 'time capsule' for the preservation of its parents' genes in hard times. Seeds have the ability to lie dormant for many years. The record for survival is held by an arctic species of lupin whose seeds were found in a frozen lemming burrow. They had been there for 1,400 years, yet they were able to germinate and grow when the surface of each seed was filed to let in water. Variability in the timing of germination is as important as it is in the hatching of eggs when the environment is fickle. Like the mosquitoes that live where the rains are unreliable, desert plants produce crops of seeds that germinate at different times. Some germinate after a splash of rain but run the risk of withering if no substantial rain follows, while others will not germinate until they have been well soaked.

Dormancy is also advantageous for seeds as they are the means by which the species is transported to new ground. As plants are immobile and are literally rooted to the ground, their seeds play an important role in the dispersal of the species. In contrast, most eggs are static, since animals are mobile and can themselves disperse – with the exception of sedentary marine animals, whose eggs float in the sea, and some land animals whose eggs are carried by other animals. Seeds are carried on the wind, in water, on the fur or feathers of animals, or even inside their bodies – having been swallowed, they are excreted at a later date with the animal's droppings. One of the best examples of the seed's ability both to survive and to take the species to new ground, is the coconut. The reason why coconut palms are such a feature of tropical islands around the world is that the 'nuts' survive prolonged immersion in salt water, drifting for thousands of kilometres before being cast ashore to germinate on new territory.

Above left: nuts, like eggs, are good to eat, but unlike eggs they can turn this drawback to advantage, because animals will help to disperse them. If this wood mouse stores its hazelnut instead of eating it on the spot, the nut will have a chance to germinate.

Jays (*below left*) bury acorns as a food supply for the winter. They remember later where they are hidden and dig them up, but some acorns are forgotten and germinate at a distance from the parent tree.

Right: these seeds of an African acacia, seen germinating after rain, will have been lying dormant during the dry season. Seeds are like eggs in many ways. They carry the developing embryo and are also the means by which the species is distributed to new habitats or preserved during adverse conditions such as drought.

CHAPTER 2
Eggs for all lifestyles

The egg of an invertebrate living in the sea – a mussel, a ragworm or a starfish, for example – is very different from that of a land-dwelling animal. The differences between the eggs of marine and freshwater animals are less obvious to the naked eye, but are still important. Three main factors account for these differences: the 'saltiness' of the sea, the dryness of the atmosphere and the time taken for the egg to develop and hatch.

The oceans were the cradle of life and compared with the land, the sea is a very favourable environment. Because there is little variation in temperature or chemical composition, marine animals and their eggs do not have the physiological stresses that face terrestrial or freshwater animals.

The majority of marine invertebrates and fishes lay small, simple eggs without shells and without much yolk. They are shed into the sea and either lie on the

sea bed or float freely but helplessly for a short time before hatching into larvae. The larvae swim and feed for a while, then change into the adult form. This system avoids the necessity of providing the egg with copious quantities of yolk. In contrast, freshwater invertebrates lay eggs which nearly always develop for a comparatively long period before hatching into juvenile animals resembling the adults in miniature. A floating larva is a liability in a river as it would be swept away and lost. Freshwater eggs, therefore, carry more yolk to support longer embryonic development and produce more robust offspring. Also, they are

Below: the freshwater crayfish lays her large-yolked eggs in November and they remain attached to her swimmerets throughout the winter. In spring, the young hatch out as miniature versions of their parents. They cling to their mother's swimmerets with their claws until their first moult. During this time they are sustained by remnants of the yolk that are stored in their bodies.

Below: small winkles, *Littorina neritoides*, cluster in damp cracks so far up the shore that the average high tide only splashes them. However, spring tides reach the winkles every two weeks, so they liberate their eggs into the sea every fortnight. Each egg is contained in a buoyant capsule. In about six days it hatches into a free-swimming larva which feeds in the surface water for about two weeks before settling on the bottom. Here it will metamorphose into the adult form and start the long crawl up the beach.

Right: a clutch of eggs laid by the great grey slug. Everything needed for the development of the next generation of slugs is stored within the protective membrane that encloses each egg.

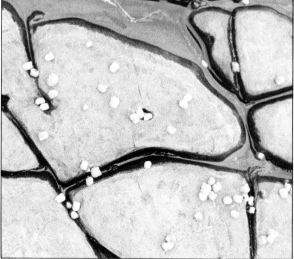

often protected by being kept in the female's body until they are ready to hatch.

The most important difference between sea and fresh water is its chemical composition. Marine invertebrates have roughly the same concentration of salts in their body fluids as the water that surrounds them. For freshwater invertebrates the problem is that the low salt concentration in the surrounding water is out of balance with their body fluids. Water is drawn into their bodies by osmosis, the movement of water through a membrane from a weak solution to a more concentrated one. Without a barrier to control the movement of water or a means of removing excess, the animals would become distended with water, but their excretory systems (the equivalent of our kidneys) bale out the excess and prevent this from happening. The eggs of freshwater invertebrates do not have excretory organs to pump out the excess fluid, but they do have physiological mechanisms for pumping salts and water across the membranes, thus maintaining the difference in salt concentration.

The eggs of freshwater animals absorb a certain amount of water after they have been laid, but a balance is soon reached. In other respects the developing egg is largely insulated from the adverse environment. The long development time results in the young animal hatching with an excretory system in working order and able to cope with the flood of fresh water that continually enters its body through its permeable skin.

The eggs of animals that live on land, or on the seashore where they are exposed by the falling tide,

need protection to prevent them from drying up. They are provided with a shell that is a container for their own watery environment, but no egg is completely waterproof. There is always a limited passage of water through the shell and uptake of water may be essential for their development. When the environment dries up, loss of water is resisted by the properties of the protoplasm of the egg, but eventually the egg will dry up if loss continues.

It is tempting to search for close links between an animal's lifestyle and the type of eggs it lays. Take for example the winkles living on European shores. The four most abundant species, although closely related, survive in very different conditions. The small winkle lives in crevices in the splash zone above the extreme high-tide mark. The rough winkle lives further down, to the mid-tide level, and the flat winkle is found lower down still and overlaps with the common winkle that reaches the spring tides' low-water level. The position of each species reflects its ability to survive out of water: the small winkle is wetted by spray only during the spring tides, while the common winkle is covered almost constantly. This suggests that winkles are demonstrating the process whereby animals became terrestrial through increasing their ability to survive desiccation and the warmth of the sun.

There is a parallel development in the winkles' reproductive habits. The common winkle, like many wholly marine animals, sheds eggs into the sea where they soon hatch into floating larvae. The flat winkle has lost the larval stage; it lays its eggs on seaweed and they hatch into baby winkles. The rough winkle goes

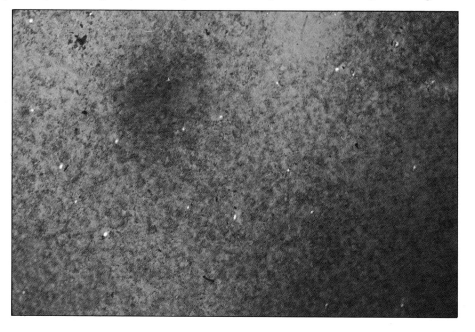

The shell of a land animal's egg is a container that supports the egg tissues and reduces water loss. It is not totally watertight as it must allow the passage of gases for respiration. Birds' eggs are provided with tiny pores for this purpose, as can be seen in this enlarged photograph of a hen's egg. Each species' egg has a total area of pores that is related to the needs of the embryo inside.

Grey sea slugs spawning in a shallow intertidal rock pool. These sea slugs are commonly found between tide marks, where they actively prey on sea anemones. Their spiral egg ribbons are surrounded by jelly which protects them from drying out should their pool empty between tides.

Below: most starfishes shed sperms into the sea and leave them to swim to the eggs, but *Archaster typicus* is unusual in that two individuals come together in a form of mating. The eggs are laid soon after fertilization and lie on the sea bed. Also unlike most other starfishes, whose eggs hatch into floating larvae, these large, yolky *Archaster typicus* eggs will hatch into baby starfishes.

further and retains its eggs in its body until they have hatched. This enables it to live high up on rocky shores and in saltmarshes where eggs would survive with difficulty. The small winkle, at the top of the shore, spoils the story and reverses the trend. At the highest winter tides it lays eggs that hatch into floating larvae. Although almost capable of surviving on land, the small winkle is still tied to the sea by its conservative egg-laying. The same restriction makes it necessary for the land crabs of the tropics to return to the sea each year to spawn.

The difficulty in relating an animal's egg-laying habits to its lifestyle is also shown by the dogwhelk, which looks rather like a winkle and lays its eggs in tough-coated, flask-shaped capsules. Several dogwhelks gather to lay their capsules together, so masses of upright yellowish 'flasks' can be found in rock crevices. The capsules presumably protect the eggs within, but it is difficult to explain how. Research has shown that the eggs sometimes dry up, while many embryos are fatally affected by changes in salt

concentration when the capsules are exposed to sun and rain at low tide. Neither do the capsules totally deter predators, since large numbers are eaten accidentally by winkles as they graze on the seaweed-covered rocks. The answer is that no defence system is perfect, but it will reduce the chances of an attack being successful. The egg capsules slow down the rate of water loss and deter some predators. One species of buckie, a kind of whelk living on the shores of Alaska, has overcome the problem of predators. Its egg capsules are readily eaten by sea urchins, but they are usually laid next to large sea anemones. The anemones catch any sea urchins coming within range, thus giving the buckie protection while possibly benefiting from the buckie's capsules acting as bait.

Another suggestion is that the egg capsules of dogwhelks are an aid to the eggs' respiration. Many winkles, whelks and limpets lay their eggs embedded in a flat strip or plate of jelly. Compared with laying the eggs in a compact mass – in which the embryos near the centre are starved of oxygen and develop slowly – a flat strip lets oxygen circulate freely to all the eggs. Egg capsules have a further advantage: the oxygen that is easily absorbed because of their shape can be transported by fluid circulating inside the capsules.

One problem for animals which have learned, so to speak, to cope with fresh water is the drying-up of their environment. They survive as eggs, or less often as adults or larvae that form resistant stages. Any sample of mud from the bottom of a dried-up pond may yield a crop of animals if placed in water. Drought can happen in any part of the world but it is a special problem for animals that live in desert pools. In the depressions, known as *playas* in North America, pans in southern Africa and *chottes* or *sebkhas* in North Africa, the ground may have been bone-dry for years, yet within a few hours of filling they spring to life and swarm with protozoans. These provide meals for the hordes of tiny crustaceans that hatch out from their eggs a day or two later.

The burst of life is short-lived because the water soon evaporates under the strong sun or drains down into the sandy soil, so the inhabitants of these pools must be able to take instant advantage of a downpour; they have to produce eggs or other resistant forms before the drought returns. For many animals eggs are then the sole repositories of the species.

The eggs either hatch quickly, producing another generation to take advantage of continuing favourable conditions, or become dormant. Some crustacean eggs start developing before entering the dormant stage and this gives them a head start when the rain returns. Once dormant, they have incredible powers of survival. Crustacean eggs have been known to hatch after lying in dry mud for as long as 15 years.

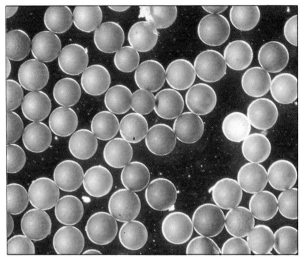

The eggs of the brine shrimp can withstand almost total desiccation, they can be frozen and they can resist many caustic chemicals. Very few other living organisms are so robust. The dried eggs shown here are concave on one side (1). In water, the eggs soon swell to become spherical (2). The minute larvae hatch after about 24 hours. At first they emerge from the eggs in neat pear-shaped sheaths (3), but soon they start to swim with jerky movements of their arms (4). The larvae feed by sieving plankton from the salt pools in which they live. Growth is rapid. When the adults pair, the male clings to the female with specially developed claspers (5). Eggs are initially sited in the tail of the female (5) but, after pairing, the eggs are transferred to a pouch (6).

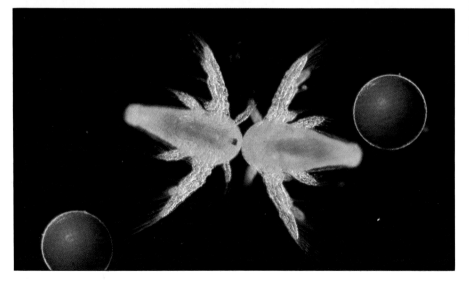

Moreover, they can tolerate extremely harsh conditions. Eggs of the tadpole shrimp (*Triops*), a common inhabitant of short-lived pools, have been known to withstand temperatures of 98 degrees centigrade for several hours, and those of the brine shrimp can be stored at minus 20 degrees centigrade. Brine shrimp eggs have a tough shell, but it is not responsible for their survival in ultra-dry conditions. The egg tissues can dry out until only two per cent of water is left, and still survive.

How these eggs survive is not fully understood. It is known that their shells are not totally waterproof, so it would seem that the tissues of the embryos within must be physiologically adapted to survive desiccation in some way, and that the eggshell is a protection against abrasion and the intense desert sun.

Too much water can be as lethal as too little. Slugs lay their eggs in crevices in the soil, where they are in contact with the film of water coating the soil particles. If the soil becomes waterlogged the eggs will die, but some insect eggs are protected by their ability to slow their growth in adverse conditions. Some are also able to avoid drowning because they have a network of gas-filled spaces under the shell, which connect with the surface through tiny pores. When the eggs are submerged, air is trapped inside.

Drowning the eggs has, in a roundabout way, been used to get rid of an insect pest, the Australian bush fly that was probably imported by European settlers. This relative of the house fly lays its eggs on the hard crust of cowpats. As cattle farming spread through Australia, so did the bush fly. Conditions were ideal because the native dung beetles which fed on marsupial droppings could not cope with cattle dung. Eventually European dung beetles were imported to cope with the dung problem; as a bonus, they also destroyed the bush fly, because their excavations in the cowpats pushes the fly eggs into the mushy centre where they drown. The result is that huge areas of Australia have been cleared of bush flies.

Amphibians

The familiar species of frogs, toads and newts lay their eggs in still, fresh water. Each egg has a coating that swells on contact with water to produce the ball of transparent jelly enclosing the egg. The jelly is 99.7 per cent water, but it is vitally important for the survival of the eggs, as it protects them against predators. It also acts like the windows of a greenhouse, letting the sun's rays through to warm the developing eggs – whose black colour is the best for absorbing heat – while at the same time preventing the warmth from escaping. The temperature of the eggs can be raised by as much as two degrees

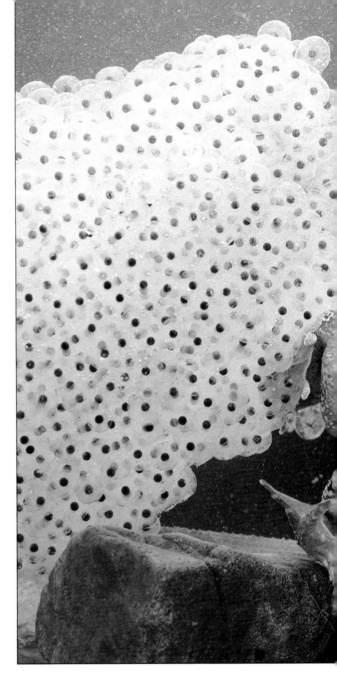

centigrade, which enables the eggs to survive and speeds development in temperate countries such as Britain where the water can be very cold during the spawning season.

Amphibians spawning in cold water, which holds plenty of dissolved oxygen, typically lay their eggs in globular masses, while species laying in warm, oxygen-deficient water produce strings or flat plates of spawn – as we have seen, this enables sufficient oxygen to reach all the eggs. Mountain torrents are cool and rich in oxygen, but the problem here is to prevent the spawn from being swept away. In these

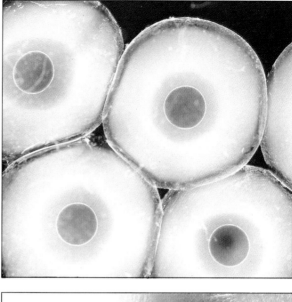

conditions amphibians lay single eggs or small masses which are anchored or hidden under stones, and some species lay in the damp vegetation beside the stream.

Within the jelly surrounding the egg there is a lining which slowly dissolves, so the advanced embryo floats within its jelly sphere. Eventually the tadpole hatches out, either by dissolving the jelly with enzymes secreted from a gland on its snout or by nibbling with its teeth. For a while it rests in a cup formed by the remains of the jelly, which continues to act as an incubator until the tadpole becomes active.

Above left: a common frog lays around 1,000 eggs at a time. Some of the spawn seen here will die, because it has been laid in shallow water and will be stranded when the level drops.

Top right: the eggs of the common frog are surrounded by tough, clear jelly. This swells by taking up water during the first few hours after being laid and forms a protective covering which acts like a greenhouse, trapping the sun's rays to warm the eggs. Warmth speeds development, giving the tadpoles an early start in spring.

Bottom right: a few frogs and toads lay their eggs out of water because they have found ways of keeping them moist. The grey tree frog of southern Africa lays its eggs in a mass of froth where they are protected from enemies. After the tadpoles have hatched out, their wriggling liquifies the froth and they drop into the pool below.

31

Whereas the fishes returned to the sea from fresh water, the amphibians have never taken this step. Many have, however, advanced further than the fish in adapting to breeding out of water. In the tropics, for example, where there is often a shortage of standing water, amphibians breed on land in damp places, among dense mats of moss or under stones and logs where moisture is held.

The tadpoles of Rattray's frog of South Africa actually drown if placed in water. The Australian corroboree toad lays its eggs in a burrow, where both parents may remain for some weeks. The embryos develop to a particular stage and then pause until a rise in water level floods the burrow. The primitive *Leiopelma* frogs of the misty mountains of New Zealand lay eggs in damp hollows and the tadpoles live on the damp ground, but other species hatch out as froglets, omitting the tadpole stage. In these cases the eggs are well provided with extra yolk to feed the embryo.

Land-based eggs are often protected by a tough capsule. Alternatively, one parent whips the jelly into a foam as the eggs are laid; the surface of the foam solidifies and becomes waterproof so that the eggs and tadpoles develop in a sort of aquarium.

Even stranger than the amphibians who lay their eggs on land are those that carry their eggs with them. In Europe there is the male midwife toad who carries his mate's spawn wrapped around his legs. He urinates on the eggs to swell the jelly as they are laid and at intervals he goes to the water to soak them. The marsupial frogs of South America carry their eggs in pouches on their backs. When about to spawn, the female usually raises her hind legs and the eggs roll down her back and into the pouch, although in one species, *Gastrotheca marsupiatum*, the male pushes the eggs into the pouch with his feet. The young are released, as tadpoles or froglets according to species, by their mother holding the mouth of the pouch open with her hind legs. Perhaps the strangest of all is the rare – or possibly extinct – gastric brooding frog of Australia. The female swallows her eggs and the tadpoles develop in her stomach. In due course the froglets emerge through her mouth.

Left: the Roman or edible snail lays its clutch of eggs in the soil. The eggs are protected by chalky shells and the young will hatch out as tiny replicas of the adult snail.

Right: tadpoles of the many species of rain frog living in the wet forests of tropical America complete their development in the egg and hatch out as froglets. In some species the tail is large and well supplied with blood vessels to act as a lung.

Reptiles

Some specialized amphibians have virtually cut free from the ties of water both for everyday life and for breeding, but they do so with difficulty. The reptiles on the other hand have made the transition, in a major evolutionary jump, to an entirely land-based life. Their bodies and eggs are much more waterproof than those of amphibians, and the embryo completes its development in the egg – no reptile goes through a larval stage. Compared with the amphibians, therefore, reptiles are independent of the environment and have the potential to take up ways of life denied to their amphibian forebears. They tolerate desert conditions more than any other vertebrate class and several reptiles – turtles, crocodiles, snakes and lizards – have taken up a marine life.

Part of this success is a result of the evolution of the egg, or more precisely its container, from the simple jelly-covered amphibian egg. The reptile egg is called 'cleidoic', meaning 'box-like', and is the forerunner of the bird egg, although the details of its construction vary between groups. The shape is usually spherical, oval or elongated, but some are pointed at both ends. Sizes range from 5 millimetres in small lizards, to 120 millimetres in pythons. The shell, which is always white, is composed of layers of fibres lying at right angles to each other. The eggshells of turtles, lizards and snakes are soft or parchment-like, but those of crocodiles, tortoises and geckos are impregnated with calcium salts and are hard like a bird's egg. In turtles and crocodiles, albumen or 'egg white' provides the embryo with an extra store of food and water. Crocodile eggs also have an airspace at one end, as in birds' eggs, but curiously this is missing in other reptiles.

The most important features of the reptile egg are the three membranes surrounding the embryo: the amnion, chorion and allantois. These act as a life-support system for the embryo, making it independent of the environment, and are a major adaptation for terrestrial life. Reptiles, birds and mammals are called amniotes, to distinguish them from amphibians and fishes which do not have these membranes. The embryo of a reptile or a bird survives on dry land because these membranes provide a water-filled bath, act as lungs for exchanging oxygen and carbon dioxide with the atmosphere, and form a reservoir for waste products. How these membranes function is described in more detail in Chapter 7.

Birds

Birds are 'glorified reptiles' and some biologists believe that they are indeed no more than reptiles with feathers. It is not surprising, therefore, that birds' eggs are similar to those of reptiles. The main interest in birds' eggs from an evolutionary point of view is the adaptation of the eggs of different species to particular lifestyles.

In both reptile and bird eggs, there is a large yolk which not only supplies the needs of the embryo but will also sustain the newly hatched animal until it can feed. In birds' eggs, after the egg cell – consisting mainly of a ball of yolk – has been fertilized, it moves down the oviduct where it collects the albumen or 'egg white' and takes on its shell. The albumen is mainly a filler for protecting the embryo, but it also supplements the yolk as a valuable source of protein and a reservoir of water. Between the albumen and the shell are two parchment-like shell membranes. When they are first formed they fit quite loosely, but during the next stage, known as 'plumping', water and salts fill out the albumen until the membranes are taut. The shell is then attached firmly onto the outer membrane, as can be seen when peeling a hard-boiled egg.

The innermost layer of albumen is thick and forms a hammock slung between the ends of the shell to keep the yolk suspended in the centre. Within it are two twisted cords called chalazae, which act as bearings to allow the yolk and embryo to rotate freely when the egg is turned by the parent bird in the nest. Chalazae are presumably unnecessary in reptile eggs because they are never turned, and some, such as those of tortoises, will not hatch if they are turned accidentally.

The shell takes 15 to 16 hours to form in a domestic hen. It consists of crystals of calcite, a form of calcium carbonate, strengthened with protein fibres which also attach it to the underlying membrane. The amount of calcium needed for each egg is about 2 grams. As no more than 25 milligrams are circulating in the blood of the hen, the balance must be mobilized from her reserves. She cannot absorb enough calcium from her food while making the egg, so she has to remove it from her bones. Birds have developed a special kind of bone, called medullary bone, which females lay down in the marrow cavity of existing bones during the weeks before laying commences and use as a source of calcium when the eggshells are being formed. Some birds even have difficulty in laying down this reserve because their

Left: the eggs of the grass snake, like those of other reptiles, are adapted for life on land. The tough shell protects the embryo and the internal membranes provide for its respiratory and excretory needs. The eggs are laid in the warmth and moisture of rotting vegetation to give them the best chance of hatching.

Right: a spur-thighed tortoise arranges her eggs with her hind feet after laying them in a pit dug in the soil. If they are to hatch successfully, she has to ensure that the clutch is properly packed and that no eggs remain exposed.

diet is deficient in calcium. Shortage of calcium affects the breeding of scavenging birds, such as vultures, which have a diet of pure meat; to overcome this problem some species switch to hunting small animals, which they swallow whole, during the breeding season. Another example of the ingenious methods used by birds to provide calcium for their eggs can be seen on the Arctic tundra, where insect-eating sandpipers swallow the bones of long-dead lemmings; as a result there is more calcium in each clutch of eggs than there is in the adult's entire skeleton.

The shell of a hen's egg is peppered with 10,000 tiny pores, just visible to the naked eye as small depressions in the shell. These pores are used for the passage of oxygen, carbon dioxide and water vapour. Unlike a lung, however, the flow of gases cannot be regulated and relies purely on diffusion through the pores. Nevertheless, the flow must be finely attuned to the needs of the embryo. As development proceeds, more oxygen is required, but the loss of water must be kept low. The eggshell's role in regulating gas flow and water loss is shown by the arrangement of pores in different eggs – quite simply, diffusion depends on the size and number of the pores. Of all the eggs studied, from the 1-gram eggs of warblers to the 1.5-kilogram ostrich eggs, the weight of the egg and hence its metabolic requirements is directly proportional to the total cross-sectional area of the pores.

Left: guillemots and kittiwakes nesting on an exposed cliff ledge. The kittiwakes make high-sided nests of mud and grass to contain their eggs, but the guillemots lay their eggs on the bare rock. To lessen the chance of falling off, guillemot eggs (*below*) are tapered so that they roll in a tight circle. By comparison, a hen's egg rolls in an almost straight line.

The four eggs in a lapwing's clutch (*right*) fit together neatly so that the incubating parent can easily cover them; they are also coloured and patterned so that they are beautifully camouflaged when left uncovered. When the eggs passed down the oviduct extra pigment was added to give the pattern of spots. By contrast, the eggs of the dunnock or hedge sparrow (*far right*) have only a single 'ground' colour.

Birds' eggs are uniform in shape when compared with the eggs of reptiles, but there are significant variations from the standard oval. Owls' eggs are noticeably more spherical than normal, while those of swifts and swallows are longer and more elliptical, although it is not obvious why this should be. Guillemots have tapered eggs, which roll in tight circles and are consequently less likely to fall off the narrow ledges where they are laid. The shells of the eggs are thickened at the pointed end which rests on the ground and bears the brunt of any movement. Many waders have eggs shaped like old-fashioned spinning tops, blunt at one end and tapering sharply at the other. The reason for this is clear: there are usually four in a clutch and, being large for the size of the parent, they fit under its body better when packed points inwards in the nest.

When birds first evolved, their eggs, like those of reptiles, must have been white, and many species still lay white eggs. Sometimes this is beneficial. For example, woodpeckers and kingfishers lay their eggs in deep holes or burrows; because they are white they show up well in the dim light and are less likely to be trodden on and smashed. Birds such as penguins, pigeons and ducks, which incubate very closely or cover the nest on leaving, also have white eggs.

The shells of coloured eggs are coated with a pigment as the eggs pass down the oviduct. Some have a single 'ground' colour – the translucent green of ducks and the beautiful deep blue of the dunnock –

while others have patterns of blotches, spots and streaks. If the pigment is added when the egg is stationary in the oviduct, the result is spots; if it is added when the egg is moving, streaks are produced.

Eggs are often coloured for camouflage. The pattern breaks up the outline of the egg in much the same way as camouflage paint disguises a tank. The colouring also matches the background – this is especially important for birds nesting on open ground. Some birds can even vary the colouring of their eggs according to where they are nesting; for example, they lay brown eggs among vegetation and blackish eggs on burnt ground.

The patterns also enable some birds to recognize their own eggs. Guillemots, for example, lay their eggs on bare cliff ledges, without any nests, so it is essential that a bird returning to the mass of breeding birds can identify its own egg. The shell of each guillemot egg has a unique combination of ground colour and blotches, and it seems that the parent bird is able to memorize this pattern.

The greatest feat of recognition is performed by the ostrich. Several females lay their eggs in one nest but only one female incubates them, with the assistance of the male. With 30 or 40 eggs in the nest, it is impossible for one bird to cover them all adequately. The incubating female pushes some of the other females' eggs to the edge of the nest, apparently being able to identify her own by the pattern of pores in the plain white shells.

CHAPTER 3

Fertilization: the start of life

Fertilization, the meeting and joining together of sperm and egg, is the starting point of life for almost all animals. From this moment there is the potential for development and, with the fusion of the genetic material from male and female parent, the blueprint for a new individual is determined. The information contained in the genes, which have been shaped by the forces of natural selection acting on ancestral animals, will now be shuffled to produce unique details of form, physiology and behaviour.

The outcome of this shuffling will determine whether the new animal will endure the stresses of its environment and survive to contribute its own share of genes to the next generation. There is more to fertilization, however, than the mixing of parental characteristics and the egg and sperm are more than passive containers of genes. The cytoplasm making up the bulk of the egg cell also contributes to its future and the sperm is needed to trigger development.

Fertilization can take place only if eggs and sperm are brought together when both are 'ripe' or fully mature. Except in some of the simplest animals, the tissues which eventually become eggs are laid down in the early embryo and develop slowly, if at all. Then, as the time of fertilization and laying approaches, there is a spurt of growth during which an egg cell outstrips all other cells in size. The mouse egg enlarges some 40 times, but the eggs of birds with their enormous stores of yolk show the greatest increase of all.

The ripening of the eggs follows a different pattern in different animals. In humans, one egg ripens and is liberated into the oviducts every month, alternately from each ovary. This is known as ovulation. Songbirds usually ovulate once a day, laying one egg 24 hours later, until their clutches are complete, while the wandering albatross ovulates and lays only one egg in alternate years. At the other end of the scale, the starfish *Asterias* sheds its 100 million eggs in one massive act of spawning.

Ripening of the eggs and ovulation in humans and some other mammals is automatic, occurring at regular intervals unless interrupted by pregnancy, but in most other animals ovulation is seasonal. The main condition for the timing is that breeding should coincide with a plentiful food supply.

Once the egg is ripe, it must be brought into contact with the fertilizing sperms. The egg is, by definition, passive and needs to be sought by the active sperms. These are essentially torpedoes with warheads of genetic material fired by the male in massive salvos with the hope that at least one will hit its target. Hundreds of millions of sperms are launched, yet the chances of any reaching their destination are often extremely slight. The wonder of fertilization is that it ever takes place against such enormous odds.

Part of the handicap is that both sperms and eggs are short-lived. The 'fertile life' of an egg, when it remains capable of fusing with a sperm and giving rise to a healthy embryo, is usually measured in hours. The eggs of humans, cattle and guinea pigs remain viable for about a day after ovulation, while those of mice, rabbits and rats live for about half this period; sea urchins' eggs, on the other hand, are viable for 40 hours. Fertilization may take place after a longer period but the embryos are likely to develop abnormally.

The fertile life of sperms is often longer, particularly in insects. The female insect usually mates only once, but the sperm she receives remains viable throughout her reproductive life. It is stored in her body and released whenever she lays a batch of eggs. A queen honeybee, for example, stores sperm from a single mating for several years. Among reptiles, diamond terrapins are known to be able to lay fertile eggs four years after mating. Bats are unusual among higher animals because they mate in autumn and the females store the sperm in the uterus, where it is nourished through hibernation until fertilization takes place in spring.

Both male and female animal must produce and release their sperms and eggs at roughly the same time and, because a sperm can swim only a few centimetres, in roughly the same place. Some method is also required to improve the chances of eggs and sperms meeting. For simple animals the answer has been the 'spawning crisis'. Many marine

Bitterlings are small freshwater fish that deposit their eggs within the mantle cavity of mussels. Here the eggs develop in a protected environment until they hatch. Some days later, the young fry swim forth via the outlet siphon of the mussel.

Some species of bitterling have long ovipositors, but the Japanese species *Rhodeus ocellatus*, shown here, has a short ovipositor which it inserts into the mussel's inlet siphon in a rapid dash, thus avoiding getting it trapped as the mussel closes. The female also has a long trailing organ which grows behind the ovipositor during the breeding season. This is a signal to the male that she is ready to lay.

Male bitterlings guard mussels during the breeding season and drive off all other fish except females that are ready to lay. The male displays to the female and, as soon as she has laid her eggs, deposits a small puff of sperm over the siphon of the mussel.

A pair of giant African millipedes mating in an embrace that may last for several hours. The genital openings are on the third body segment and the male possesses an intromittent organ for placing sperm in the female's body.

Below: the strange mating posture of a pair of common blue damselflies. In damselflies and dragonflies, the male transfers sperms from the sex organs at the tip of his abdomen to a storage organ on the front end of his abdomen. Then he grasps the female's head or thorax with a pair of claspers and she swings her abdomen forward, bringing her sex organs into contact with his storage organ. The male remains attached to the female while she lays her eggs, to prevent other males mating with her.

Snails are hermaphrodite, each individual having both male and female sex organs. These Roman snails are about to fertilize each other. As a mutual stimulus the snails inject sharp 'love-darts' into each other's bodies. These contain a chemical that acts as a stimulant to mating. The snails then exchange packets of sperms called spermatophores, which can be stored for long periods so that fertilization and egg-laying may take place quite a while after mating.

invertebrates, such as mussels and sea urchins, synchronize their spawning and the water becomes thick with eggs and sperms. The crisis is started by one sea urchin, for example, shedding its gametes; almost immediately a wave of spawning spreads through its neighbours, who are probably stimulated by a chemical messenger – a pheromone – that is shed with the spawn. For a mass spawning to take place, all the eggs and sperms must be 'ripe' at the same time; their development will have been triggered earlier by cues from the environment, such as changes in day length, temperature, moon or tide.

The classic story of co-ordination comes from the tropical Pacific, where the palolo worm is famous for its mass spawning on four nights of the year. The palolo worm is a bristleworm, a relative of the earthworms, that lives in burrows in coral reefs. As spawning time approaches, the rear half of the body undergoes a cataclysmic change: the internal organs degenerate and this half becomes little more than a bag of either male or female reproductive organs with two rows of paddles for swimming. When ripe, and with the right stimulus, it breaks from the rest of the body, which remains in the burrow, and propels itself to the surface, where eggs and sperms are released.

By spawning at the surface the gametes are concentrated in one place, but they are also concentrated in time. So precise is the spawning of the palolo worm that Fijian islanders, for whom the worms are a great delicacy, either cooked or raw, can predict exactly when to launch their boats to scoop up the writhing masses. First they watch for certain plants to come into flower, and then they look for the moon resting on the horizon at dawn. Ten days later, on the first two nights of the last quarter of the October moon, there is *Mbalolo lailai* – the little palolo. This is followed exactly a month later by *Mbalolo levu* – the big palolo.

In 1984 there came the discovery of massive synchronized spawning among corals on the Great Barrier Reef of Australia. To date, more than one hundred species have been found to spawn within a few hours of each other on a few nights of the year.

Synchronized spawning greatly increases the chances of eggs being fertilized but it is still very wasteful. If the chances of fertilization are improved still further, the female can lay fewer eggs containing more reserves of yolk. It may then be worth her while to protect them by laying them in a nest. To ensure fertilization of large, valuable eggs, the male must approach and shed his sperm near the eggs as they are being laid. This happens with crayfish, where the female carries her eggs under her abdomen, and with fishes such as salmon and sticklebacks which lay their eggs in nests with the male in close attendance. Toads and frogs and such fishes as Siamese fighting fishes go one step further. Their eggs are shed into water but fertilization is enhanced by the male embracing the female and shedding his sperms at the same time.

The main significance of this meeting of the sexes is that they pair off. A male can now court a specific female to ensure that he fathers her offspring and the female can choose which male is to be that father. The scene is now set for males to compete with each other and court females for the privilege of mating and transmitting their genes to the next generation.

The male crayfish mates with the female by turning her over and embracing her while he sheds a packet of sperms, the spermatophore, onto the underside of her abdomen where it sticks to her swimming legs or swimmerets. In due course she lays her eggs, which also stick to the swimmerets, and fertilization takes place. This idea has been developed in crabs, such as the common shore crab: using specialized limbs the male inserts his sperms into the female's oviduct, where they are stored until the eggs are shed.

The next stage is for the eggs to be retained within the female's body until they are fertilized. Internal fertilization is a huge advance in the development of sexual reproduction, and is one that has evolved several times in the animal kingdom. It confers several advantages beyond reducing wastage of eggs and sperms. By retaining the eggs in her body, the female can increase their protection. She can also add extra food supplies and a protective shell after fertilization has taken place.

Internal fertilization requires an act of copulation in which male and female come into intimate contact for the transfer of sperms. This is usually preceded by the pair courting to confirm the identity of the species, fertility of the female and suitability of the male. Sperms are usually introduced into the female by the male's intromittent organ, the penis or equivalent. Sharks have a pair of claspers, so called because they were once thought to be for holding the female; and the male guppy (a popular aquarium fish) has a gonopodium, a long tube formed from part of the anal fin. Insects and snails have intromittent organs, but birds, with a few exceptions such as the ducks and the ostrich, do not; they press their cloacas – combined genital and excretory organs – together. Newts, unlike frogs and toads, practise internal fertilization but mate at a distance. The male deposits a spermatophore and the female, following closely, picks it up in her cloaca.

Platynereis megalops, a bristleworm related to the palolo worm, copulates in so strange a way that some zoologists have been unable to believe that it occurs. The worms swarm at the surface of the sea like the palolo worm, but internal fertilization is necessary because the eggs lose their viability after only 40 seconds' contact with sea water. The male wraps his body around the female, pushes the rear end of his body into her mouth and ejects his sperms. As the female is little more than a bag of eggs and the intestinal wall has broken down, fertilization immediately takes place and the eggs are shed.

There are two bizarre variations on the theme of copulation. The male spider secretes a spermatophore, picks it up in his pedipalps, the small limbs between the front legs, and transfers it to the female's genital opening. He has, of course, first disarmed her by courtship or he might even have tied her down with silk; usually the female is larger than the male and will not hesitate to attack or even eat him if the opportunity arises. Squid, cuttlefish and octopuses use a modified tentacle, the hectocotylus, to deposit a spermatophore in the female's body. A relative of the octopuses, the argonaut or paper nautilus, leaves its detached tentacle in the female; when first discovered by the Italian zoologist Stefano delle Chiaje, in 1872, it was believed to be a parasitic worm and given the name *Hectocotylus*, meaning 'arm of a hundred suckers'.

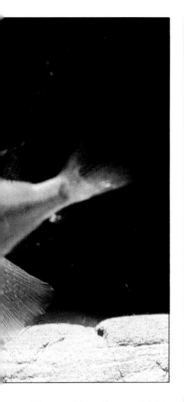

The mouthbrooders are cichlid fishes which carry their eggs and young in their mouths. In this species from Lake Malawi the eggs are carried by the female. The male entices her to a flat stone which he has picked clean of algae. She lays her eggs on the stone, then immediately turns and picks them up in her mouth. The male swims alongside and displays the 'egg dummies' on his anal fin. The female mistakes these spots for more eggs and attempts to seize them, but gets a mouthful of sperms instead, which fertilize the eggs in her mouth.

Even with the advantage of copulation placing sperms within the female's reproductive system, fertilization of the egg is not easily accomplished. In mammals, of the hundreds of millions of sperms sent on their way, only a few reach the eggs waiting in the oviduct a few hours later. These survivors will have found their target by chance: as torpedoes, they are very simple and have no guiding mechanism. Only in some relatives of the corals and the Portuguese man-o'-war is there any evidence of a chemical attracting sperms to eggs.

Having reached the unfertilized egg, the sperm must penetrate the surrounding membrane before entering the tissue of the egg. It can then deliver its load of genetic material, which fuses with that of the egg to create the blueprint of the new individual.

The mechanisms of penetration and fusion have traditionally been studied with the aid of the transparent and easily obtainable eggs of sea urchins and other marine invertebrates, although recent research into the problems of infertility has revealed a great deal of information about the fusion of the human sperm and egg. In general, details vary from species to species, partly through differences in the structure of the egg and sperm, but the main sequence of events remains the same.

As soon as the random swimming of a sperm has brought it into contact with a mature unfertilized egg, it must burrow through the surrounding matter to the surface. The egg may be coated in a thick jelly as in sea urchins or amphibians, or a cluster of cells as in mammals. Most of the tiny fraction of sperms that have got this far fail in the attempt to cross this barrier – up to 90 per cent in sea urchins. Once through this obstacle, the next stage is to penetrate the vitelline coat, a stout envelope enclosing the egg, called the *zona pellucida* in the mammal egg. Again, many sperms fail in their mission at this point, but the successful ones fuse with the egg membrane and complete the process of fertilization.

As the sperm touches the vitelline coat, its head, the acrosome, secretes enzymes from a 'pepperpot' arrangement of holes and dissolves a passage through to the egg. Sperms of insects and fishes lack an acrosome, but they penetrate to the egg through tiny openings called micropyles. Whichever method is used, an infinitesimally small fraction of the sperms originally launched by the male reach the ultimate goal and dock with the egg membrane enclosing the protoplasm.

Normally fusion can only take place between an egg and sperm of the same species, because the egg 'recognizes' the identity of the sperm by chemical means. Fusion takes only a few seconds and is followed by another sequence of events. The space between the egg membrane and the vitelline coat swells and the coat itself develops into a hard 'fertilization membrane'. One function of these changes is thought to be to keep excess sperms out of the egg. In sea urchin eggs the electrical charge that is found in all cell membranes changes drastically and excess sperms which had previously attached themselves are thrown off. A very quick response is thought to be necessary because a synchronized spawning 'crisis' can lead to a swarm of sperms gathering around the egg. In mammals, on the other hand, where few sperms will have fought their way up the female's reproductive system and struggled through the surrounding tissues to the egg, there are unlikely to be any surplus sperms. Occasionally the blockage is incomplete and extra sperms do penetrate the egg. When this happens several sperm nuclei fuse with the egg nucleus and an abnormal embryo is produced. Other animals, such as birds, reptiles and insects, employ a different strategy. Several sperms can penetrate the egg but only one nucleus is allowed to fuse with the egg nucleus.

Once inside the egg, the sperm nucleus moves into the cytoplasm to meet the egg nucleus. Both enlarge, becoming clearly visible under the microscope, and come into contact in the centre of the egg cell. The two groups of chromosomes now intermingle and pair up. This is the closing event of fertilization and the start of the complex processes of development in which the single cell of the newly created individual is transformed eventually into a living animal.

Eggs without males

In 1935, a Russian zoologist, Ilya Darevsky, published the results of collecting 5,000 specimens of the Caucasian rock lizard. He had not found a single male and he put forward the theory that these lizards were reproducing by eggs that had not been fertilized. If this was true, he realized, it would be the first known case of parthenogenesis or virgin birth among vertebrate animals.

Proof of the theory came a few years later with studies of whiptail lizards living in the arid country in the south-west of the United States and in northern Mexico. As lizards can store sperm in their oviducts for months or even years, it was necessary to follow the history of known individuals very carefully. Young female lizards were kept in captivity and encouraged to lay eggs by being given the right food and conditions. Only females hatched out from these eggs and they in turn laid more eggs until hundreds of females of several generations had been produced without the appearance of a single male.

Probably one quarter of the 40 species of whiptail

lizards are parthenogenetic and they have arisen as a result of hybridization. For instance, the parthenogenetic New Mexico whiptail lizard *Cnemidophorus neomexicanus* is a hybrid of the little striped whiptail lizard *C.inornatus* and the western whiptail lizard *C.tigris*. Normally, hybrids are sterile, like the mule produced by a horse and donkey. In the whiptail lizards, the chromosomes in the developing egg double an extra time so that the egg eventually has pairs of chromosomes rather than a single set. It is therefore in the condition that it would attain if it had been fertilized and received chromosomes from a sperm.

Aphids also produce diploid eggs but the mechanism is rather different. Instead of the second

Huge numbers of aphids build up in spring and summer because they give birth to live young by parthenogenesis. These females are wingless but, at intervals, winged females appear and fly in search of another plant on which to start a new colony; no male is needed so an infestation can start with a single aphid. In autumn males appear and the females produce fertilized eggs which survive the winter.

doubling of the chromosomes, as in the whiptail lizards, the second reduction is omitted, but the result is the same. The queen honeybee employs yet another system. As we have seen, she is fertilized only once, on her mating flight, and the sperm is stored in her body. Thereafter, she lays unfertilized, parthenogenetic eggs, which hatch into male drones, or fertilized eggs which develop into female workers or queens. Chromosomes are reduced to single sets when the eggs ripen, so fertilized, female eggs become diploid, while drone eggs remain haploid.

Parthenogenesis has the same advantages and disadvantages as asexual reproduction. It produces clones, with identical genetic composition. Without any variation, they are susceptible to changes in the environment and epidemics of disease. The advantages are a potential doubling of the breeding rate, because the whole population lays eggs, and a single individual can establish a new population if it reaches a previously uninhabited area.

The advantages of parthenogenesis in this respect are well illustrated by the aphids, which alternate between parthenogenetic and fully sexual generations. Only one parthenogenetic aphid need alight on a plant to give rise to the masses of black fly or green fly which cloak roses, beans, sugar beet and other plants. Without the need to mate, generation time is very short.

Parthenogenesis is recorded in about 1,000 species, including aphids, bees, wasps, ants, grasshoppers, stick insects, crustaceans and rotifers (tiny animals that live in fresh water and damp soil). Among vertebrates, there are some parthenogenetic salamanders and, under artificial conditions, turkeys and chickens. In these birds meiosis has gone wrong and they have laid diploid eggs. These hatch into male birds that on very rare occasions have proved to be fertile.

It will be remembered that the function of sperm is twofold. As well as providing genetic material, it stimulates the development of the egg. In parthenogenesis, both functions are lacking. Experimentally, unfertilized eggs have been stimulated into development by a variety of means. Frog embryos have been grown by pricking the egg with a needle dipped in fresh frog's blood. Cold, heat, electric currents and chemicals have also been used to stimulate development, apparently by acting on the egg membrane. How the stimulating function is replaced in nature is not known except in the case of a parasitic wasp, *Habrobracon*, whose eggs are stimulated mechanically when squeezed down the oviduct.

There is also the case of the Amazon molly, a fish which is a natural hybrid of two other species of molly. Virtually the entire population is female. Amazon mollies reproduce by mating with males of one of the parent species, the sailfin molly. It seems that their sperms penetrate the eggs and stimulate development mechanically but do not contribute genetically.

The role of the male

When naturalists examined the crowds of tropical butterflies that they frequently saw drinking by the banks of streams, they were surprised to find that they were all males. This was something of a puzzle until it transpired that the butterflies were gathering where large mammals had urinated on the ground and they were drinking the salt-rich fluids from the mammals' urine. The purple emperor, one of Europe's most spectacular butterflies, has a similar, and to human eyes unsavoury, habit. The female is rarely seen because she lives high in the canopy of oak woods, but males are sometimes spotted in woodland glades when they come down to drink from rotten, suppurating corpses of animals or from puddles of liquid manure. The probable reason for this behaviour is that male butterflies need a good supply of sodium salts, and perhaps also potassium and calcium, to secrete in their spermatophores. After copulation the salts are absorbed by the female's reproductive organs for use in egg production.

Male and female supply an equal amount of genetic material to the fertilized egg, but the female normally supplies the egg with all the yolk and other materials that will nourish and protect it. The butterflies show how a male can ease the burden of egg production by supplying his mate with extra nourishment.

Transferring salts in the spermatophore is unique to these butterflies, but stranger still is the blister beetle's transfer of cantharidin, the irritating substance known as 'Spanish fly'. Cantharidin is a defence against predation and female blister beetles extract it from the spermatophore to enhance their own defences – they may also add it to the eggs to make them more unpalatable. Other insects transfer proteins in the spermatophore. For example, female crickets eject the empty spermatophore and eat it, so that the proteins can be used to build the eggs, and female katydids (an American grasshopper named after the rhythm of the male's song) choose to mate with the largest males who will have the largest spermatophores, thus increasing their egg production.

A more common way of supplying nutrients to the female is to feed her. Robber flies, for instance, present the female with prey when courting, and many male birds feed their mates both during courtship and while the eggs are being incubated. 'Courtship feeding', as it is termed for birds, is partly a ritual for maintaining the bond between the pair but, especially in small birds such as tits and finches, it is also of vital assistance to the female when she is producing the eggs. The total clutch can weigh more than her body and, even though she does not carry

The male purple emperor butterfly sips perspiration from a human hand. Along with other sources of salt-rich fluids, such as rotting carcases, this provides the male with extra salts which he passes to the female in his spermatophores to help her produce eggs.

A cock fantail pigeon displays and calls to the hen. Both sight and sound of the male stimulate the development of the female's eggs.

them all at once, the weight of a single egg can make flying laborious and hinder her ability to go foraging.

The female common tern gives up fishing completely for the few days before egg-laying starts. The male, who until then has been only supplementing her diet with occasional courtship feeding, provides all her food while the yolk, albumen and shell are being laid down in her three eggs. His efficiency at performing this duty affects the pair's subsequent success as parents. The more food he brings, the larger the eggs the female will lay, and large eggs hatch into large chicks which have a better chance of survival.

Male spiders and praying mantises are famous for making the ultimate sacrifice for their offspring. They are often eaten by the female after or even during mating and their bodies nourish the eggs that they have just fertilized. This is not as invariable as is sometimes suggested, since male spiders, which are generally much smaller than the females, exercise caution when approaching a prospective mate. They signal their identity and intentions and take care to avoid the female's jaws. As a result, some spiders may live to mate several times, perhaps with different females. The same is true for mantises. They often avoid being eaten, but sexual cannibalism is usual at the end of the summer when food is becoming scarce and the females are hungry.

Courtship is a tricky business for the smaller male spider who must persuade his mate not to eat him. This male green orb spider courts the female by stroking her with his front legs. He has already lost one leg from a previous mating attempt. When the female is receptive he transfers sperm in his pedipalps and, unless he can retreat quickly, he is caught and eaten.

49

CHAPTER 4
Launching the eggs

Eggs are the female's investment in the future. They contain her genes and she has expended energy and her body reserves such as proteins and calcium in their manufacture, so it is in her interest to provide them with the best chance of survival, and perhaps to care for the young when they have hatched out. Assisting the survival of the eggs, at its simplest, may be no more than laying them in the most favourable situation; at its most complex it may involve a high degree of parental care in the form of nest-building, incubation behaviour and retaining the eggs within the female's body.

Many invertebrate animals lay their eggs in a packet – a cocoon or pod – which gives greater protection than the shell alone. For the earthworm, the cocoon also plays an important part in the fertilization of the eggs. Earthworms are hermaphrodite; each individual is both male and female (the term recalls Hermaphroditus, the half-man half-woman offspring of the gods Hermes and Aphrodite). When a pair of earthworms copulate, each receives the sperms of the other and stores them in special organs called spermathecas. About a day after mating, both worms begin to lay batches of eggs. The cocoon is

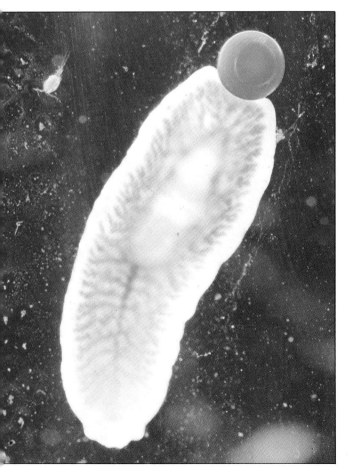

manufactured by the clitellum or saddle, the enlarged segment about one third of the way down the worm's body. It consists of a tube of fibrous material into which albumen is secreted to sustain the hatchling worms when they first emerge. The earthworm slowly wriggles backwards out of the cocoon, which picks up eggs from the opening of the oviducts and sperms from the opening of the spermatheca. As soon as the worm has passed through the 'tube', the ends of the cocoon seal into points; within the cocoon the eggs are fertilized by the sperms. Similarly, spiders protect their clutches of eggs by wrapping them in silk. The yellowish cocoons can often be seen nestling in corners of rooms that are not kept thoroughly dusted.

Snails and slugs lay their eggs in masses in the soil, often without any extra protection, but many freshwater and marine snails surround their eggs with a stiff jelly or capsule. Pond snails stick their jelly masses to the leaves of water plants, but if they are kept in an aquarium they will attach them to the glass. After a storm, the round masses of whelk egg capsules are tossed onto the shore, as are those of cuttlefish which look like bunches of grapes. The female

cuttlefish lays her eggs in batches of 20 or 30. Each egg is coated with a latex-like solution as it is laid and coloured with a squirt of ink for camouflage. The solution is drawn out into a long stalk before it sets, then several capsules are entwined to form the mass.

Some insects lay their eggs in a capsule called an ootheca. As each batch is laid, it is cemented together with a fluid secreted by glands in the female's reproductive system that hardens in the air or as a result of chemical reaction. Praying mantises fashion their oothecas from a gummy substance which is frothed to a crisp meringue; it protects the eggs but also allows them to breathe. The cockroach, on the other hand, has to perform an elaborate system of moulding to provide the eggs with an air passage. As each egg is laid, a tiny 'horned' structure in the female's reproductive tract is aligned with the breathing pore in the eggshell and acts as a mould for the cement-like substance which fills the space in between the eggs in their capsule. When it has hardened, the 'horns' are withdrawn, leaving two air tunnels connecting each egg with the outside atmosphere.

Far left: a freshwater flatworm lays its eggs in a spherical capsule which will be glued to the underside of a stone or leaf. The eggs are surrounded by a mass of yolk cells that will form an initial food supply for the young worms.

Left: a tiny spider with its outsize egg cocoon. The tough case of the cocoon protects the brood until the spiderlings emerge.

Right: the eggs of a pond snail near the time of hatching. The baby snails can be clearly seen inside.

An ichneumon fly searches for the grub of a wood-boring beetle, trying to detect its characteristic odour with her antennae. Any likely hole or minute crack in the surface of the wood is carefully probed with her long ovipositor. By rotating her body back and forth she can force this horsehair-thin organ deep into the gallery of the beetle larva. After the egg has been laid, the ovipositor is withdrawn and returned to its protective sheath.

A particular feature of many insects, that is not often found in other animals, is the possession of an ovipositor, an egg-laying tube which enables the female to place her eggs with accuracy in otherwise inaccessible places. (Bees and wasps have turned the ovipositor into a sting for delivering venom.) Among the robber flies there is considerable variation in the structure of the ovipositor and its use. Some species have a simple tube and merely scatter their eggs on the ground, while others have a flap hanging over the opening of the tube to deflect the eggs into the protection of crevices. *Philonicus* robber flies, who live on the sea shore, use a simple ovipositor to bury their eggs in soft sand; they are also equipped with two long bristles on the abdomen which are used as a broom to sweep over the traces of their activity.

Other robber flies lay their eggs in plants. *Machimus* has a simple ovipositor for pushing eggs into open flower heads; *Dysmachus* has a blade-like ovipositor for piercing grass seed-heads and *Eutolmus* has a stiff, flat 'knife' for slitting plant stems.

Of all ovipositors, none is so marvellously constructed and used as that of the ichneumon fly *Rhyssa persuasoria*. 'Ichneumon' means 'tracker' and the flies in this group do indeed track down the larvae of other insects in which to lay their eggs. *Rhyssa* has a 35-millimetre ovipositor, longer than the rest of her body, and she uses it to lay eggs on the larvae of a woodwasp, the horntail *Urocerus gigas*. The horntail also has an impressive ovipositor which is used to drill into the trunks of conifer trees, where its eggs are deposited. Despite the apparent safety of living deep in timber, a horntail grub can be located by *Rhyssa* by the smell of its droppings deep in its tunnel. She detaches her hair-thin, flexible ovipositor from its protective sheath, drills down to the grub and senses

A large egg-laying organ is a conspicuous feature of the bush-crickets. The fine up-curved and blade-like ovipositor of the female speckled bush-cricket is edged with minute teeth for sawing into the stems of plants, where the eggs are deposited singly.

Eggs of the great green grasshopper are laid in soil. The long, slightly down-curved ovipositor is not a digging implement but will be inserted by the female into suitable crevices in the ground.

through her ovipositor whether it is in the correct condition – if it is, she lays an egg in the tunnel. The hatchling larva attaches itself to the horntail and sucks its blood. But even the ichneumon larva is not safe. There are other species of ichneumon that search for the minute shafts left by *Rhyssa*, insert their own ovipositors and lay their eggs. When the larvae hatch, they kill the *Rhyssa* larvae, then eat the horntail grubs.

Another almost incredible example of the selection of an egg-laying site is that of the fly *Stomoxys ochrosoma*, whose larvae live in the nests of the ferocious driver ants of tropical Africa. The larvae and adults of many insects make a living as scavengers in the nests of social insects – bees, wasps, ants and termites – where there is a plentiful supply of waste material and dead insects on which to feed. The problem is how to get into the nest without being attacked. A *Stomoxys* female searches for a column of nomadic ants, finds a worker which is not carrying anything and quickly drops her clutch of 20 eggs just in front of it. The ant's instinct is to carry the eggs of its own species, so it readily accepts those of the fly; the eggs are transported back to the nest – through the front door, so to speak – where they hatch.

Stomoxys takes care to keep clear of the driver ants by dropping her eggs from above, but a chalcid wasp, *Methoca chalcoides*, positively courts disaster. Chalcid wasps are parasites that lay their eggs on the bodies of other insects, paralyzing them first, so that the larvae have a ready store of fresh food. *Methoca* searches for the burrow of a predatory tiger-beetle larva and allows herself to be seized in its jaws. Before the beetle larva can penetrate *Methoca*'s hard body, she stings and paralyzes it. Then she lays an egg on its helpless body and, on leaving the burrow, blocks up the entrance with soil.

53

Clustering and scattering

Putting all your eggs in one basket is said to be risky. However, the alternative of an egg in each pocket is likely to lead to at least some breakages, whereas a basketful, handled carefully, will reach home intact. There is a fine balance between these arguments in nature as is shown by the egg-laying behaviour of butterflies and moths. Even closely related species of butterflies differ in their egg-laying strategies.

The small white butterfly *Pieris rapae*, which lays on cabbage and nasturtium plants, dots its eggs one at a time on the upper sides of the leaves with apparently little care in choosing each site. Its close relative, the large white *P. brassicae*, spends much more time selecting a good site on the underside of a leaf, where it lays a clutch of a dozen or more eggs. The chances are that predators and parasites, such as chalcid wasps, will find at least a few of the small white's eggs, but with large whites, it is all or nothing: either the clutch is found and destroyed or it stays hidden and all the eggs hatch. Evidence that both these strategies are successful can be seen in vegetable gardens where cabbages often have a skeletal appearance caused by the ravages of caterpillars of both species. In the evolution of these contrasting strategies, the various considerations in favour of one or the other carry different weight depending on the species' lifestyle, and only a detailed analysis of their natural history will reveal why a particular strategy has been preferred.

The advantages of laying eggs in clusters or scattering them singly extend beyond the advantages afforded to the eggs themselves. The interests of larvae and adults are also served by egg-laying strategies. Cryptically coloured butterfly and moth caterpillars almost always arise from eggs that have been laid singly. Clearly, the effectiveness of camouflage would be considerably lessened if many such caterpillars hatched out together. Conversely, species with distasteful caterpillars often lay eggs in clusters and the caterpillars remain together to reinforce the protective effect of warning coloration.

This distinction explains the egg-laying behaviour of the two species of white butterflies. The solitary larvae of the small white are a perfect leaf green and difficult to see, whereas the gregarious caterpillars of the large white are distinctively marked and distasteful to birds.

Clustering has other advantages for caterpillars. Warmth generated by their metabolic processes promotes growth, so caterpillars in groups grow more quickly than individuals that have become separated and are living alone. Groups of caterpillars can also effectively protect themselves from predators and parasites by constructing a communal silk tent, as is made by the larvae of the small tortoiseshell butterfly.

Below left: the small white butterfly lays its eggs singly on the upper surface of leaves, whereas the closely related large white butterfly lays small clutches of eggs on the lower surface. Small white caterpillars are attractive to predators; to counteract this they are camouflaged and live singly. On the other hand, large white caterpillars, *below right*, are distasteful, boldly marked and gregarious. Each of these opposing strategies has its advantages but there is a fine balance between them.

Right: lackey moth caterpillars on a silken web near the egg mass from which they have hatched. Laying eggs together saves the female moth time and energy, while living together gives the caterpillars protection from predators.

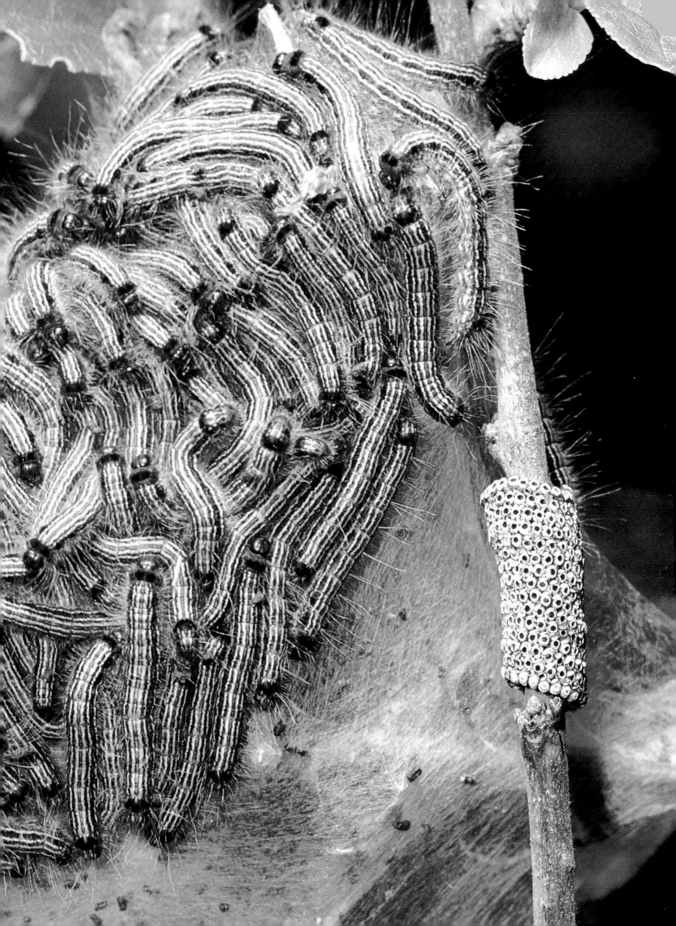

There are also advantages for adult butterflies in laying egg clusters. In countries that have unpredictable summers, egg-laying is often held up by cool, windy weather; as soon as there is a short spell of sunshine the butterflies have to make the most of it, laying their eggs as quickly as possible. Cluster-laying scores in this situation because flying time is reduced, and it is also more successful when the host plants are patchily distributed; a butterfly runs a greater risk of being caught and eaten before all its eggs are laid if it has to spend time searching for a separate site for each egg.

On the other hand, the advantage of laying eggs singly is that both eggs and caterpillars are less likely to be found by predators and parasites. Moreover, the caterpillars will not face competition from each other. In some moths, notably the eggars, egg-scattering is taken to the extreme. The females fly low over heath or moor, broadcasting their eggs as they go. This strategy only pays off because the caterpillars thrive on a wide range of food plants, so their chances of finding food are high. Furthermore, the female moth needs only enough energy reserves for one flight.

Energy and egg numbers are intimately linked and if egg-laying uses up a large amount of energy, egg numbers must be correspondingly reduced. Flying in search of many sites and laying eggs singly will use more energy than laying clusters. Various moths have evolved to a stage where egg-laying requires virtually no energy and all available resources are put into egg production. The female vapourer moth needs only the energy to burst from her pupal skin and cocoon

and await the arrival of a male. Once fertilized, the eggs are deposited on the outside of the cocoon and the female dies, exhausted. The female common bagworm moth of America has an even more limited existence. She remains inside the pupal case where she is mated, then she dies without laying her eggs. When they hatch, the caterpillars make their way out of her body and the cocoon. Wings would be an extravagance in these circumstances, and the females of both these species have none.

The contrasting strategies of clustered and scattered egg-laying provide another example of the importance of variety in evolution. Clearly, no single strategy has an overriding advantage over others. However, the somewhat complex behaviour involved in carefully selecting a site on which to deposit all the eggs has evolved in relatively few species and, for most moths and butterflies, scattering appears to be the best strategy.

The time to lay

Throughout the temperate lands of the world, spring is the time for breeding. Activity is most obvious among the birds whose outpourings of song and eye-catching courtship rituals indicate that nests will soon contain clutches of eggs, but at the same time ponds are filling with frogspawn, newly sprouted leaves are being dotted with the eggs of butterflies, and less conspicuous animals are laying eggs unnoticed in nooks and crannies. It is obvious that breeding should take place when the weather becomes warmer and food more plentiful, but the stimuli that set in

Above right: sea hares live for much of the year offshore among the kelp. In the breeding season they congregate in large numbers in the shallows to mate. Frequently they copulate in chains; being hermaphrodite, each sea hare in the chain acts as a male for the one in front and a female for the one behind. Their strings of orange or pink eggs are deposited around weed. A medium-sized string may contain 140,000 eggs, which hatch into tiny swimming larvae.

Left: a female vapourer moth waiting on her cocoon for a male to arrive. The female has no wings and she lays her eggs on the cocoon itself. Her immobility saves energy which she can divert into egg production.

motion the sequence of events that leads up to the laying of an egg are less clear. Indeed, the mechanism controlling the timing of breeding in the majority of animals is still unknown, although many of the details have been worked out for birds.

The first stage in the production of an egg takes place long before the animal is even born. The tissues which will become the eggs are laid down in the early embryo, but they do not develop until the animal becomes sexually mature. As the time of fertilization and laying approaches, there is a spurt of growth in the eggs and they outstrip all other cells in size. Then, when the eggs are mature, they are released in time to be met by the sperms. This sequence of events is controlled by a number of 'triggers' which ensure that not only are the males and females brought together, but also that the eggs are laid at the most appropriate season. Essentially this means that breeding must be timed so that there is sufficient food available both for the production of eggs and to feed the young when they hatch, to give them a good start in life. Where

the environment is uniformly favourable throughout the year, as in parts of the tropics, breeding may take place at any time, but where the climate is seasonal and there is only a short period of summer plenty, breeding must be more strictly timed.

On the arctic tundra, for example, there is a short but very productive summer during which the ground is carpeted with flowers and the air is filled with insects. Snow buntings, geese and other birds fly north to breed during this season of plenty, but they must waste no time. Eggs have to be laid as soon as the snow melts if the young birds are to be strong enough to accompany their parents south before the snow returns. A hard winter or a late spring thaw will delay nesting, with fatal results for the young birds at the end of the summer; in extreme circumstances, the birds may not even attempt to lay any eggs.

If eggs are to be laid and hatched at the most suitable time, their development must commence well in advance. The tawny owl lays its eggs in March or April when mice and voles are in short supply, whereas the owlets learn to hunt for themselves in August when prey is abundant. Therefore egg production, which must start several weeks before laying begins, cannot be triggered by an abundance of prey, although this will determine the survival of the owlets in the summer. There must be some other trigger, perhaps the increasing length of the days after midwinter, that sets off the egg's development at the beginning of the year. Some migrant birds even start the physiological preparations for laying when still thousands of kilometres away in their winter home, which bears no resemblance to the environment in which they will rear their young.

So although food supply cannot be the trigger that brings these birds into breeding condition, it is the 'ultimate factor' which, over the course of the species' evolution, has led individuals to breed at the most suitable time.

The ultimate relationship between food supply and breeding season is well illustrated by three European falcons. The kestrel feeds its young on small rodents which are easiest to find before the vegetation has grown, so it nests early, in April. The hobby lays in late May or early June and feeds its chicks on the young of swallows and martins and on large insects such as dragonflies, all of which are on the wing in late summer. Finally, Eleanora's falcon, a Mediterranean species, does not lay its eggs until the second half of July, because it feeds its young on the multitudes of small birds that pass through the area in autumn on their way south.

To get the timing of the breeding season right, animals need 'proximate factors' that will bring them into breeding condition and start the development of the eggs, so that laying will take place at the time which has been 'decided' by the 'ultimate factor'. In those species which have been studied, there is usually more than one proximate factor at work and together they give a good forecast of when the most favourable period for breeding will occur.

Chaffinches and other birds kept in uniform laboratory conditions and given no clue about the passage of the seasons, still come into breeding condition at about the correct time of year. Their rhythm of behaviour is controlled by an 'internal clock' that gives them a sense of time, but the 'clock' is not completely accurate and in natural circumstances needs resetting at intervals by rhythmic stimuli from the environment. For example, daylight is a good timing device for triggering the breeding cycle because the length of the day changes through the year in a reliable and predictable fashion. However, slavish reliance on constant timing devices could lead to disaster. If birds, for example, built their nests on a fixed calendar date, a late spring or a cold spell would ruin their attempts to breed. A second trigger that relates to current conditions is needed, so that the birds can take advantage of an early spell of good weather or delay breeding in a late spring.

Temperature is a good indicator of the state of the environment and birds can respond to warm weather by building nests and laying eggs within a few days of

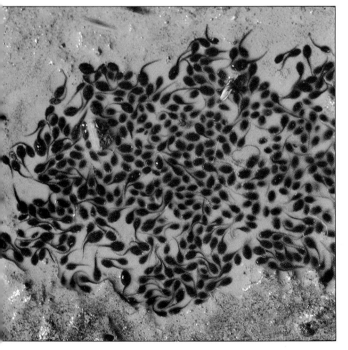

The blue tit has adapted its breeding season so that the eggs are laid in time for the maximum amount of caterpillars to be available for feeding the nestlings. If the spring is mild, the female lays her eggs earlier and takes advantage of an early flush of caterpillars.

The striped pyxie toad of East Africa spawns in the wet season; for the tadpoles, it is a race against time to grow up before the pools left by the rains dry up. They become trapped in animal footprints where the last of the water gathers and will survive so long as there is wet mud. If the ground dries too soon, the bodies of the tadpoles form a crust that will act as a fertilizer for algae during the next rainy season, thus benefiting the next generation.

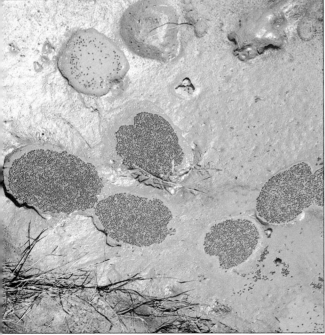

The date of spawning of the common frog varies from pond to pond. This is due to variations in local climate which affect the development of the eggs in the females and the growth of algae on which the tadpoles will feed.

the change. The difference between an early and a late spring changes the egg-laying dates of great tits nesting in southern England by six weeks. As the emergence of the caterpillars on which they feed their young is similarly affected by the weather, responding to temperature changes is clearly a great advantage.

In parts of the world where the seasons do not follow in orderly succession, it is even more necessary to take advantage of a change in the environment. In arid countries, for instance, rainfall can be irregular, or even if there is a regular rainy season its start may be delayed. Animals must take advantage of the arrival of the rains and start to breed with the minimum of delay. Plants respond to the dampening of the soil, producing fresh leaves, flowers and seeds which provide food for animals, but they soon wither as the ground dries again. The birds of these regions remain in breeding condition for long periods, so as soon as the rains come, nesting and laying can commence. For the quelea, a sparrow-like bird living in the African savannahs, the sight of grass turning green is the trigger for breeding, and older birds even respond to the sound and feel of falling rain.

The maturing of the eggs may be automatic, as in humans for instance, where one egg ripens and is liberated into the oviducts every month, or as in dogs, mice and many other mammals, where eggs mature and are shed in a regular 'oestrus' cycle. Courtship and nest building are the final triggers in birds. The 'billing and cooing' of the male ring dove stimulates development of the female's eggs; and doves in solitary confinement have even been induced to lay by the sight of their own reflection in a mirror. Canaries, and presumably other birds, react to the feel of the nest through a heightened sensitivity of the skin of the breast and belly. As a canary builds its nest into a close-fitting cup, it is automatically induced to ovulate.

Laying together

One of the many tourist attractions in California, and maybe one of the most unexpected, is the spawning of the grunion, a small saltwater fish that would pass unnoticed were it not for the fact that hundreds of thousands of them flap and leap out of the breakers together, to lay their eggs on the sandy beaches. This mass spawning makes the 'grunion run' a good tourist spectacle because it can be timed to within a few minutes. The grunions appear on the first three or four nights following both the full and new moon from February to September, starting precisely at high tide. The shallow water becomes crowded with jostling fish, then a few pioneers throw themselves onto the sand and the rest follow until there is a silvery carpet covering the beach.

As soon as the grunions come ashore, the females wriggle into the wet sand to lay their eggs and the males follow to fertilize them. Two weeks' later, at the next high spring tide, vibrations from waves breaking on the shore stimulate the eggs to hatch and the young grunions swim away.

One purpose of this remarkable behaviour is to bring the males and females together at the appropriate time and place, but it has the additional function of reducing the danger to the adult fish, the eggs and the young. Predation is reduced in two ways. The number of adults or eggs can simply swamp the predators, creating a surfeit that they cannot take advantage of. This must happen off the coast of North America when shoals of the squid *Loligo pealei* gather

Below: lesser flamingos nesting on a lake in Kenya depend on the lake attaining a certain depth if nesting is to proceed. If it is too deep they cannot build their nests; if it is too shallow the nests are easily robbed. Nesting in a dense colony gives them their only protection against some predators such as the marabou stork (*above*). The marabou feeds mainly on offal and needs prey such as nestling flamingos to give it the calcium needed for egg production.

to spawn in spring. The migration from deep water is heralded by the warming of the sea and egg-laying is triggered by the lengthening days. The result is mass spawning and there are reports of squid eggs covering three square kilometres of sea bed. Predators will no doubt eat huge numbers of them, but masses will be left. Secondly, there is safety in a crowd. Each squid that lays its eggs in the mass will have a better chance of leaving surviving offspring because the larger the mass, the greater the chance that someone else's eggs will be the unlucky ones to be eaten.

Another aspect of synchronized laying is that there is less time for predators to eat the eggs. The Fijians cannot have many palolo worm feasts (see page 41) because the worms come to the surface on only two occasions in the year. On these few nights there is a limit to what can be scooped up, so many escape.

The same principles are at work in bird colonies. The penguins in the Antarctic have few enemies but their eggs and chicks are stolen by skuas, close cousins of the gulls, which make a habit of supplementing their fish diet by preying on other sea birds. The penguin colonies are saved from complete devastation because nearly all the eggs are laid within the space of a few days. The eggs are most vulnerable shortly after they have been laid, before the parents have settled firmly into incubating them, and the chicks are vulnerable when they first hatch out. By laying in synchrony, the colony's eggs and chicks are vulnerable for only a short time. For a few days the skuas find easy meals, but there is a limit to the number that they can take each day. When eggs and chicks become harder to steal, they have to hunt elsewhere.

Choosing a site

As soon as her eggs are fully developed, a female is seized by the urge to lay them, but first she must find the right place. When a common 'cabbage white' butterfly is ready to lay, her reaction to colours changes. Until then she has been attracted to blue and yellow – the colours of the flowers from which she sips nectar – but now she prefers green and starts to search for leaves on which to lay her eggs. This is pure instinct, embedded in the genes and programmed to be switched on at the appropriate moment by a signal, probably from the eggs. Having found some leaves, the butterfly then identifies the right kind by recognizing the scent of the mustard oils that give the cabbage family its familiar odour, and chooses those of the best quality. In a field of cabbages some will grow better than others and the butterflies select healthy plants that will provide the best nourishment for their caterpillars. The choice is made on the basis of colour: well-nourished plants have deep green leaves. This is unfortunate for farmers and gardeners as fertilizing a crop of cabbages will actually attract butterflies. Tortoiseshell butterflies, on the other hand, look for pale, undernourished stinging nettles, because they contain fewer harmful toxins than healthy plants.

For females of all species the most urgent instinct is to ensure that the eggs and hatching young are provided with the right physical conditions. Burying eggs in the soil is a simple method of guarding against drying up and maintaining them at an equable temperature, but the needs of the eggs and larvae may be more precise. The mosquito *Armigeres* lays her

Choosing the right spot in which to lay eggs is crucial to the survival of the species. This is especially important when all the eggs are laid at once. The female horsefly, *Tabanus (left)*, may spend half an hour investigating a suitable pool – testing the water and trying out innumerable sites – before settling down to lay her eggs on a rush overhanging the water. After ten days the eggs darken as the larvae develop inside. On hatching, the larvae simply drop into the water below.

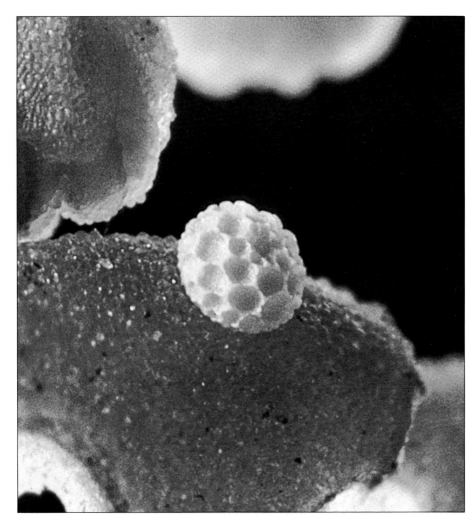

The beautifully sculptured egg of a small copper butterfly is stuck to a sorrel or dock leaf. The female butterfly has to identify the right kind of leaf because, if the eggs are laid on the wrong plant, the caterpillar's chances of growing up are impaired.

eggs on her own legs and flies in search of water. She then dips her legs into the surface and the eggs float away.

Another vital requirement is to ensure that the emerging young will have a ready supply of food. Many insects provide both amenities together: they insert their eggs directly into the plants or animals that the larvae will later eat. Some parents go to extreme lengths to provide food. For example, dung beetles collect balls of dung and the burying beetles inter dead animals for the future nourishment of their offspring.

The placing of the eggs may require elaborate behaviour. The ash bark beetle is one of the many beetles that lay their eggs under the bark of trees. They are known in the United States as 'engraver beetles', because their burrows are revealed as patterns of shallow depressions when the bark is stripped away. Preparation for egg-laying is a joint effort by both sexes. First the beetles bore into the

Left: a garden snail laying one of a clutch of eggs in a hole it has dug in the soil. It must choose soil with the correct amount of moisture, otherwise the eggs may be killed by waterlogging or desiccation.

Right: the eggs of the great grey slug are laid in a very damp spot, usually beneath a rotting log. Unlike snail eggs, slug eggs do not have a shell containing calcium and require continuous high humidity if they are to survive.

bark, leaving a small oval hole. Then, after mating in an enlarged chamber, the female excavates a broad tunnel about four centimetres long. At intervals she excavates a short side chamber and lays an egg in it. When the grubs hatch out they are not only safe from most enemies but are also surrounded by food. They eat their way through the wood at right angles to their mother's tunnel, to leave the characteristic pattern on both timber and bark.

Sometimes, when good sites are in short supply, overcrowding can lead to the destruction of nests and eggs, as can be seen on sandy tropical beaches where turtles come ashore to lay their eggs. When late arrivals excavate their nesting pits, they destroy clutches of eggs that had been laid earlier.

Overcrowding continues to be a problem if, unlike baby turtles that disperse to the sea, the young require a source of food at the nesting site. Tadpoles, for example, may turn to cannibalism, while caterpillars which do not move from the plant where the eggs were laid can die of starvation. The *Cactoblastis* moth, which was introduced to rid Australia of the prickly pear cactus that was overrunning thousands of hectares of countryside, illustrates this point well. The moths select the greenest, and hence most nourishing, cacti on which to lay their eggs. As a result some plants receive so many eggs that the caterpillars kill them and starve to death. However,

the species survives because some of the moths are less selective and lay their eggs on yellowish, less healthy plants. It is this ability to destroy prickly pears while leaving just enough to preserve their own species that has made *Cactoblastis* so valuable in the fight against the cactus pest.

Cactoblastis moths are attracted to the best egg-laying sites so they suffer from overcrowding, but some butterflies reject even their caterpillars' preferred food plants if they detect the presence of eggs and larvae of their own species. The giant birdwing butterflies of New Guinea search for laying sites on *Aristolochia* vines by starting at the bottom and working upwards. If they find an egg of another birdwing on a leaf, they move on. A female large white butterfly can spot another butterfly's eggs as she approaches a cabbage leaf. Even if the eggs are brushed off, she will still avoid laying on the leaf because, as she flutters over its surface, she picks up a scent left behind by the eggs. There is a fascinating twist to this story. Somehow, by the process of natural selection, some tropical vines have evolved dummy eggs, in the form of raised yellow lumps, or stipules (the appendages at the base of many kinds of leaf) that resemble caterpillars. These can fool visiting butterflies into thinking that another butterfly has got there first, although the stratagem only works among those species that select a breeding place by sight.

Compared with a caterpillar feeding on a plant's foliage, the larva of an insect living in the egg of another insect has a very restricted food supply and sharing it with another can be fatal. It is hard to believe that a moth or beetle egg – about the size of a grain of sand – could support another insect, but there are a number of tiny chalcid wasps that lay their eggs in those of other minute insects. One kind, *Trichogramma*, has been closely studied. When the female finds another insect's egg, she walks over it and explores it with her antennae. The host egg has to be of a suitable size and shape and must not already contain a *Trichogramma* egg. If the scent of another *Trichogramma* female is clinging to it, it is immediately rejected. However, if rain has washed the scent off, the second female proceeds to drive her ovipositor into the egg, then immediately withdraws it when the ovipositor's sense organs tell her that there is already another *Trichogramma* egg inside. Sometimes the wasp is unable to find an unoccupied egg and she is forced to lay in one that already contains a *Trichogramma* egg. In this case she chooses the largest egg she can find, with the greatest supply of food, so that there is a better chance of her larva growing alongside another.

Galls

Galls are abnormal growths which can be found on many plants, especially on oak and willow trees. Some are caused by insects but others are made by fungi, eelworms and mites. Insect galls are mostly the work of gall-wasps, gall-midges and saw-flies; they provide food and shelter for the insect eggs and larvae developing inside.

'Robin's pincushion', seen on the stems of wild roses, is one of the most familiar galls. It is named after Robin Goodfellow, the woodland sprite of folklore, but it is also called the bedeguar gall, a name that may be derived from the Persian word for 'wind-borne' in the belief that galls arrived mysteriously on the wind. Their real origin lies in the activities of a tiny gall-wasp, *Diplolepis rosae*, which lays its eggs in the leaf buds of roses in spring. When the larvae hatch out they somehow stimulate the tissues of the plant to grow abnormally into the Robin's pincushion, a hairy ball rather like a clump of moss. In the following spring, adult insects emerge, but they may not be *Diplolepis* because a variety of insects take up residence in the gall. A second gall-wasp takes advantage of *Diplolepis* by laying its eggs in the developing gall and sharing the gall tissue, and both species are attacked by egg-parasitizing chalcid wasps, related to *Trichogramma*. These chalcids, in turn, are attacked by more insect parasites.

In all insect galls, the female lays her eggs in the growing tissues, either under the bark of a stem or in a sprouting bud. The galls are well-structured, not formless growths like tumours, and each gall insect produces its own characteristic gall. But how the pattern of growth is distorted into a nursery for insect larvae instead of developing into normal plant tissue remains a mystery, despite considerable research. One suggestion is that the insect or its egg alters part of the genetic blueprint in the plant tissues so that they grow to the gall's specifications. However, at the moment this is only surmise.

Some gall-wasps have an alternation of generations between sexual and asexual states, each associated with a different kind of gall. *Neuroterus quercusbaccarum* lays eggs during the summer on the leaves of oak trees. These develop in galls, which are at first 'hairy'. The larvae feed within the safety of the galls. In October the galls fall to the ground and the larvae finish their development. The adults emerge in spring as females and lay parthenogenetic (unfertilized) eggs on oak catkins. The larvae from these eggs form spherical currant galls and the adults which emerge in summer are both males and females. They mate and the cycle begins again with eggs being laid on oak leaves.

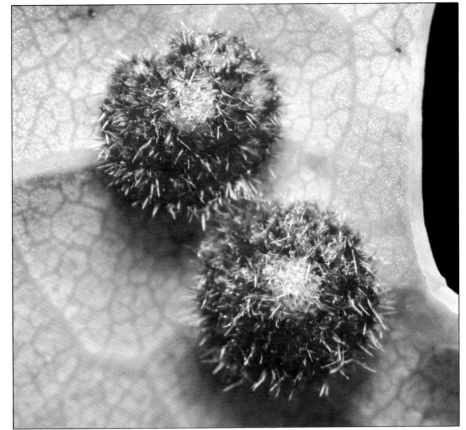

CHAPTER 5
Nests: the search for security

A song thrush has been building her nest in the garden. She has spent several days collecting grasses, leaves, twigs and mud, carrying them into the recesses of a hedge where she has worked them into a bulky cup wedged among the branches. This image of a bird's nest as a cosy home for eggs and nestlings colours our idea of what a nest should be like. It is a container for the eggs; here they will be protected from the weather and marauding predators and kept warm by the parent bird. The common birds of garden and countryside are not the only ones to build such nests; the pattern is followed by eagles, herons, cormorants, albatrosses and many others.

Building a cup of twigs and leaves is almost a characteristic of the birds, yet birds' nests come in many shapes and some only deserve to be called nests because they are places where the eggs are laid and hatched. If a nest is defined merely as a place where eggs are laid and hatched, it becomes difficult to decide which other animals are nest-builders. The engraver beetles excavate burrows in trees and lay their eggs in them. Are these burrows any different in principle from the burrows dug in river banks by kingfishers? We do not usually talk of beetles' nests, but this is what they are. Also, it should be remembered that mice and various other mammals make nests, both for their young and for the adults to sleep in.

Nests, for our purposes, are egg containers which are constructed from materials rather than from body secretions (for example, cocoons and insects' oothecas). But here again, there are awkward exceptions. The nests of the cave swiftlet, which are used for making birds' nest soup, consist of almost pure hardened saliva! There is therefore no easy definition of a nest that can be applied throughout the animal kingdom. Nests for holding eggs are built by almost all birds, the two egg-laying mammals or monotremes (the echidnas and the platypus), many reptiles, a few amphibians and several fishes. Among the invertebrates, nest-building is almost wholly confined to the insects. However, not all insects gather materials for their nests. The female millipede

works her own excrement into a sort of mortar to make a simple nest under a stone or log, or in a hollow in the soil, and some species incorporate silk or saliva into the nest so that it is more of a cocoon.

Insects

Nesting reaches a high peak in the insects with the structures built by the social insects – the bees, wasps and ants which are related to each other, and the termites which are near relatives of the cockroaches. These massive and often intricate structures are homes for colonies of hundreds or thousands of individuals, most of whom are sexless workers who do not breed themselves but devote their lives to caring for the eggs and larvae and for their 'queen', who is little more than an egg-laying machine. The ultimate object of the whole structure is to be a production line for eggs.

Not all the bees and wasps are social; some live solitary lives but show the first signs of sociality because the females lay their eggs in nests and provide food for the larvae. From here it is only a short step to continuing to supply food to the growing larvae and then to the adult young remaining in the nest; they help as 'workers' with the rearing of their younger brothers and sisters.

Some of the solitary bees and wasps excavate burrows in which they lay their eggs and supply them with nectar and pollen or prey paralyzed by their stings, but others fashion more elaborate nests. Once in a while, a newspaper will carry a story about a house being attacked by wasps and collapsing. Reading between these sensational lines, the report undoubtedly refers to the work of mason wasps. These are solitary wasps which normally make their

A long-tailed tit brings a feather to her nest. She spends a great deal of time and trouble gathering hundreds of feathers to line the nest, but the effort is repaid by the considerable insulation the lining provides, which helps to keep the eggs and nestlings warm and so saves energy.

Left: a mason wasp brings a ball of mud to make her nest. This species uses mud to fashion cells in a natural crevice. Each cell is stocked with a paralyzed caterpillar and a single egg is laid in it.

Below: the papery cells of the nest of a tree wasp are nurseries for the eggs and larvae. Several of these cells have just been vacated by a brood of young queens and drones; one already contains a new egg.

nests in earth banks but occasionally burrow deep into the mortar between the bricks of houses. They are only strong enough to penetrate old-fashioned lime mortar, so this habit will become rarer as old houses disappear. The wasps may bring water from a nearby puddle to soften the mortar, and some species use the excavated grains of sand to build a 'chimney' around the entrance to keep out rain and predators.

The nest of the potter wasp requires even more skill. As the insect's name suggests, it makes a little 'pot' from fine soil moistened with water. The pot is stocked with caterpillars and an egg is suspended inside by a thread, like a light bulb from a ceiling. When the larva hatches it remains attached to the eggshell for a time, so that it is not injured by its paralyzed but still twitching prey.

Bees and wasps lay their eggs in hexagonal cells in the comb, which bees make from wax secreted from their bodies and wasps make from paper manufactured by chewing wood and mixing it into a paste with saliva. Bees provision each cell with pollen and honey for the larva to feed on, but wasps regurgitate food for their larvae. Ants and termites have no combs and the eggs are continuously tended by workers who inspect and lick them, probably to keep them free of moulds.

At the other end of the scale there are nests that appear to be little more than special egg-laying places which provide food and shelter. The difference lies in the behaviour of the female, which includes nest-building rather than merely egg-laying.

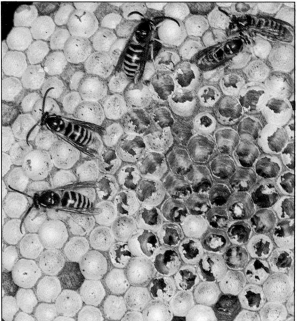

An examination of the foliage of many trees will reveal leaves that have been rolled into compact cylinders. These are the work of leaf-roller weevils and each leaf contains one egg or more, or grubs which feed on the inner coils of the roll. Before rolling a leaf, the aspen weevil takes out a small chunk to test that it is suitable; only young, fresh leaves are selected. She then walks to the leaf stalk or petiole

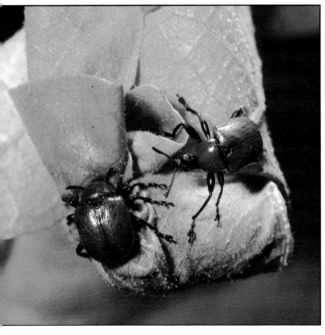

The male oak-roller weevil stands guard while the female prepares a leaf for rolling. First she cuts through the blade of the leaf using her sharp-edged foreleg (1). The leaf then hangs by its midrib and the female flexes it with her forelegs to make it pliable (2). When the leaf is suitably soft an egg is deposited on its surface, near the tip, and the leaf is tightly rolled (3). The roll is secured by a clever tuck. When sectioned, the egg can be seen near the centre of the roll, surrounded by a closely packed food supply for the emerging larva (4).

and bites a hole in it. This makes the leaf wilt and easier to roll. The weevil returns to one side of the leaf and proceeds to perforate it with many small holes which makes this part of the leaf even softer and easier to work. Next she straddles the edge and, hooking her claws into the surface on both sides, draws her legs in to curl the edge over and tucks it in with her long snout. Then she steps sideways, transferring her grip, and draws her legs in again, continuing the roll, until half the leaf is coiled. At one point she stops work, bites a small slit in the leaf and, turning around, lays an egg in it and rolls the leaf up. The second half of the leaf is rolled up in the same way and the coils are glued down. Sometimes a second leaf is added to the roll to give the grubs extra food and some weevils finish by severing the petiole completely so that the leaf falls to the ground.

The roller weevils show considerable dexterity in rolling up their leaves. It is tempting to think that they are being quite clever in making the leaves easier to work by puncturing them so that they leak sap and wilt, but it is quite simply the process of natural selection that has resulted in the evolution of such beautifully adapted instinctive behaviour.

Fishes

The majority of fishes scatter their eggs in the water. Ocean fishes, especially, leave them to float at the mercy of the currents, while coastal fishes lay dense eggs that stick together on the bottom and thus escape being swept around. Freshwater fishes often have heavy, sticky eggs and it is in fresh waters and, to a lesser extent, coastal waters, that fish nests are made, partly at least to prevent the eggs being swept away.

Trout and salmon hide their clutches of eggs in the gravel beds of fast-running, well-oxygenated streams. These may be so shallow that the backs of the fish are barely covered. The clear water gives an excellent opportunity to watch their behaviour, although spawning takes place in winter so it is a somewhat chilly exercise.

Each female selects an area of gravel called a redd where she will excavate a series of saucer-shaped nests. Turning on her side, she 'cuts' the gravel by convulsively flapping her tail to lift the gravel, which is carried downstream. Once in a while she tests the hollow with her anal fin, then continues working until she can fit her body in neatly. The male, who has been waiting nearby, darts alongside her and fertilizes her eggs as she sheds them into the nest. Up to two hundred eggs are laid in a batch, looking when newly laid like small pearls, with 'an iridescent bloom like a ripe Hamburg grape' as Frank Buckland, the Victorian fishery expert, put it.

Spawning is the work of a second and as soon as it has been accomplished the female moves slightly upstream to start cutting the next nest. The gravel from this nest is swept down to cover the eggs in the first. The eggs are now protected from the current and many predators, while being well aerated by a good flow of water.

The male three-spined stickleback is in sole charge of the nest, eggs and young. Standing on his head (*top left*), he excavates a shallow pit in the sand over which he builds a nest of weed. When he has courted a female he leads her to the nest (*top centre*) and urges her to enter (*top right*) and lay her eggs (*bottom left*). As soon as she has left, he enters and fertilizes the eggs (*bottom centre*). Thereafter, he is kept busy aerating the eggs by fanning with his fins; he must also defend the nest – and this includes removing any water snails that come too close (*bottom right*).

Although the nests of trout and salmon are simply constructed, the site is chosen with care. The gravel must have pebbles large enough to allow free passage of water, but some finer material is needed to consolidate it. Such conditions are found in the 'bars' or ridges that form at the downstream end of pools in fast-flowing streams. Part of the main flow of water over the bar drains down through the gravel with sufficient force to sweep away all the silt and sand that could clog the gravel and harm the eggs.

By contrast with trout and salmon, sticklebacks nest in quieter stretches of water (even in the still water of a pond or aquarium), where there is a fine sandy bed. Aeration is provided by the male who, as is often the case with fish, is the one who tends the nest and eggs. As the breeding season approaches, the male three-spined stickleback, familiar to generations of children as the 'tiddler', becomes brightly coloured, with red on his underside, blue eyes and a greenish back. With this striking livery, he stakes out a territory against other male sticklebacks and, when secure in his ownership, commences to build his nest.

Standing on his head, he takes mouthfuls of sand and spits them out until he has dug a shallow pit. The pit is usually located among water weed, or seaweed for the marine sticklebacks, and the stickleback constructs a tube of fine pieces of weed glued together with a sticky secretion from his kidneys. When the nest is finished, the male entices females to lay their eggs in it. On spotting a female, he swims excitedly in a 'zig-zag' dance that leads her towards the nest. He indicates it by pointing at the entrance with his head and when she enters he butts her and trembles to stimulate her spawning. After spawning, she departs. The male now enters and fertilizes the eggs, which take five to 12 days to develop, depending on water temperature. During this time the male keeps guard and fans the nest with his fins to drive a stream of water over the eggs to prevent silt from settling on them.

Reptiles

In 1978 an amateur palaeontologist working in western Montana made a discovery which shed new light on our ideas of how the dinosaurs lived. She found a fossil nest which contained not only fragments of eggshells but the complete skeletons of baby dinosaurs, of a type called hadrosaurs. Dinosaur nests had been found before, but at this site the remains had not been scattered and the scene of 80 million years ago could be reconstructed with reasonable accuracy.

Apart from providing clues that infant hadrosaurs remained in the nest and were therefore probably guarded by their parents, the nests showed how the embryos of these dinosaurs survived even though the eggs were partly buried in soil (and may have been covered with vegetation). The hadrosaurs' nests were mounds of mud with a central bowl in which the 20-centimetre-long eggs stood vertically on their pointed ends. The upright position with a space between each egg allowed the air to circulate freely, so the embryos could receive enough oxygen and get rid of quantities of carbon dioxide. Also, the shells were sculptured with patterns of ridges and bumps which would have held the soil away from the breathing pores, preventing them from becoming clogged.

The smaller eggs of modern reptiles do not have such a problem with air flow, but the female must ensure that they are laid where temperature and moisture are suitable. The leathery-shelled eggs of snakes take up water from the surroundings but also lose it if the ground dries out, while the hard-shelled eggs of turtles and crocodiles are relatively impermeable. Many reptiles lay their eggs in moist soil or under boulders and logs where it is humid, or urinate on the nest to provide the necessary moisture.

Left: a leathery turtle digging her nest on a sandy beach by night. She excavates a large trench that will accommodate her body, and then uses her hind flippers to dig a deeper pit that will receive the clutch of eggs. The depth ensures that the temperature will not fluctuate widely and gives the eggs protection from predators.

Right: the nest of an Arctic tern is the barest scrape in the sand with the pebbles and shells removed. When the sitting bird leaves the nest it is difficult to find the well-camouflaged eggs on the expanse of beach.

Far right: the wandering albatross drags earth and vegetation to its nest site and piles it up into a large mound. The single, large egg is retained in a bowl lined with grass.

To provide heat for development the nest is sometimes located in sun-warmed sand or in rotting vegetation which generates heat. However, the temperature should not fluctuate too much and burying the eggs some distance below the surface evens out the difference between day and night temperatures.

The female Hermann's tortoise, one of the species kept as pets, chooses her nest site on the basis of temperature, although it is not known how she does this. Once satisfied with her choice she laboriously scoops a hole with her back legs, reaching down as far as she can stretch to scrape alternately with each foot. The completed nest cavity is sock-shaped and the eggs are laid into the heel, leaving the toe as an air-space when the hole is refilled. The space may help supply the eggs with a good circulation of air and it is known to be essential in providing room for the baby tortoises when they hatch out.

The most elaborate reptile nest is that of the American alligator, which in effect builds a large compost heap whose heat of decomposition incubates the eggs. The alligator lives in swamps and the female constructs her nest on drier ground by clearing a space some three metres across. To achieve this she crushes the vegetation with her body or cuts it down with her teeth. She then piles up the broken plants, makes a hollow in the middle with her hind feet, refills it with more mud and vegetation and lays her eggs in the middle. Finally, the mound is raised with more mud and vegetation and moulded into a hillock one metre high.

Birds

An essential difference between reptiles and birds is that birds are warm-blooded and the embryos' body temperature has to be maintained throughout their development if the eggs are to hatch successfully. The habit of incubating their eggs in insulated nests gave the early birds an advantage over their reptilian predecessors. Warm-blooded parents can keep the eggs warm at night even in cool climates. Thus the birds have been much more successful than the reptiles at spreading out from the tropics. As nest-building became more sophisticated, birds were able to lay their eggs in safe places out of the reach of predators. Indeed, one of the advantages of incubation is that the eggs require a shorter development time and their vulnerability is reduced because there is less time for predators to find them.

Every kind of bird has its own nest, built to individual specifications and adapted to the ecology and habits of the species. There are no fossil nests to show the habits of ancestral birds and, with the enormous variety of modern nests, it is hard to trace the course of the evolution of birds' nests, especially as closely related birds can have very different nests. Compare, for example, the long burrow of a sand martin and the mud cup of a swallow. Nevertheless, it is possible to see a pattern in nest-building that may show how it evolved.

One suggestion is that birds originally sat on their eggs as a means of protecting them from predators – especially the small nocturnal mammals that were

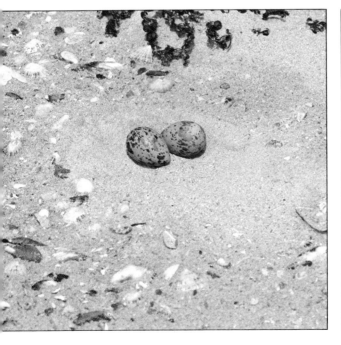

Left: the chaffinch's nest
consists of a bowl of grass, moss
and other materials worked
into a felt and lined with
feathers.

Right: the weavers exhibit the
peak of nest-building skill with
their ability to knot grass stems
into a light but strong fabric.
This male red-headed weaver
uses the foundation of his nest
as a display platform and
females will assess his
suitability as a mate by its
quality.

evolving at the same time – and that incubating them
was a later refinement that evolved when birds
became warm-blooded. Either way, scraping a
shallow saucer in the soil would help to keep the eggs
together and many modern ground-nesting birds,
such as waders, gulls or grouse, still lay their eggs in
scrapes.

A scrape is made by squatting on the ground while
scraping with the feet and pushing with the breast as
the bird slowly rotates. The combined actions
excavate and mould the scrape. From this beginning,
more material was added to make a mound of
vegetation which held the eggs snugly and securely.
For this, the birds used their beaks, just as modern
geese and albatrosses create piles of plants and soil by
reaching out from the nest site and dragging
vegetation towards them until they are sitting on a
substantial heap. The saucer for holding the eggs is
made by pulling material onto the rim and patting and
teasing it into place.

Cormorants, gannets and grebes have to search
further afield for nesting material and carry it to the
site where it is laid in position. Once the birds had
acquired the ability to carry nesting material and to
work it into a solid mass, they were all set to take up
nesting in trees where they would be safe from the

many predators living on the ground. The safety of a
tree nest is illustrated by the mourning dove of the
American prairies which nests both in trees and on
the ground. Its tree nests are twice as successful as its
ground nests.

The typical cup-shaped nest perched in the
branches seems to show a skill beyond the capabilities
of an animal that has only a beak for manipulating the
materials, but the flexibility of a bird's neck gives it
great dexterity. The nest looks as if it has been woven
but the basic technique is much the same as the
ground-nesting bird's scraping and pushing. These
actions align the stems and grasses and work them
into a strong felt, while the beak is used for lifting
material onto the rim and teasing it into place. An
additional skill required by a tree-nesting bird is the
ability to find a suitable site. This seems to be a matter
of instinct, but the equally important business of
laying a secure foundation is more a matter of trial
and error or luck, especially for birds whose nests are
made of stout sticks wedged in a fork. They use a
'fetch and drop' technique and, by the time they have
got a foundation started, sticks that have failed to
lodge litter the ground under the tree. Smaller birds
start by looping grasses around a twig or stem or by
making a pad of sticky cobwebs.

The next stage in the evolution of the nest is to build a roof which will give extra protection from the weather and predators. Protection from predators seems to be the more important reason because roofed nests are less common in temperate regions than in the tropics, where nest predators such as snakes and mongooses are much more abundant. The long-tailed tits of Europe, northern Asia and North America are exceptions. They make exquisite nests of mosses, which are lined with a thick layer of feathers and covered on the outside with lichens and cobwebs. Yet these nests are not as elaborate as those of the Cape penduline tit, which have an entrance that can be opened and closed. Beneath the real entrance there is a conspicuous false entrance and dummy egg chamber which helps to persuade predators that the nest is empty.

The variety of covered nests that occur throughout the tropics is enormous. There are the two-chambered nests of sand and dung made by the South American ovenbirds, the 30-centimetre ball of thorny twigs with a long entrance tunnel made by the spinetails, the 70-centimetre-deep nest of sticks of the aptly named firewood gatherer of Argentina, and the puzzlingly outsize nest of the African hammerkop or hammer-headed stork. Other members of the stork family make do with simple platforms of sticks, but the hammerkop spends two months fashioning sticks and mud into a single-roomed penthouse with one-metre-thick walls and roof that will bear the weight of a man. The Xhosa tribe believed that the nest had three rooms, the nest chamber being equipped with a wooden pillow and the others being used as larders. The true reason for building such a large nest is, however, a mystery.

The hammerkop's efforts may seem excessive, but the nests of several other birds, including the many species of weaverbirds which live mainly in the African savannahs and woodlands, and the orioles, oropendolas and caciques of tropical America, are even more elaborate. Weaverbirds tear slender strips from the leaves of grasses and palms and weave them into strong but light, domed nests. The village weaver starts its nest by winding a strip around a twig and then threading the loose end through the coils to secure it. More strips are woven in and the globe shape of the nest is gradually built up. Some weavers' nests hang from a woven rope that makes them very difficult to attack, and Cassin's weaver of the African rain forests adds a 60-centimetre vertical entrance tube as an extra precaution. But the most spectacular of all nests is that of the sociable weaver that lives in the deserts of south-western Africa. Each colony of these birds builds a communal nest in a single tree. Under the thickly thatched dome, which may be

seven metres across, there will be about 60 individual nest chambers. These are occupied all the year round and the larger the nest the more even the temperature will be during the cold winter.

The alternative to building a nest is to lay the eggs in a cavity. This can be a natural cavity, such as the tree hole used by tits, starlings, owls and parrots, or one excavated by the birds, such as the burrows dug by petrels, shearwaters, kingfishers and sand martins or the woodpeckers' holes in trees. These usually provide sufficient insulation and protection for a nest to be dispensed with, but the tits still make a cup of moss and grass lined with feathers, and the pardalotes of Australia build a domed nest within a cavity.

A small number of birds make no nest at all. Guillemots lay their eggs on bare rock ledges and fairy terns lay theirs on branches of trees. Owls and many falcons use the abandoned nests of other birds, while emperor and king penguins carry their single eggs balanced on their feet and walk around with them. Even more acrobatic is the palm swift which glues its egg with saliva to a palm frond, then sits on it for the incubation period.

Cuckoos

The cuckoo's fine reputation as a harbinger of spring is spoiled by the common knowledge of its lazy breeding habits. It avoids the toil and industry of building a nest, incubating eggs, rearing and feeding chicks by forcing its eggs into the care of other birds. Its offspring oust their foster brothers and sisters from the nest and so gain the undivided attention of their foster parents. To human eyes this is mean treatment of innocent birds, but from a biological viewpoint brood parasitism is an interesting solution to the problem of rearing offspring with the minimum of effort, and the European cuckoo is only one of many animals, including insects and one of the catfishes, that have adopted the habit.

The simplest form of brood parasitism is seen in birds that lay their eggs in the nests of others of their own species. This is called dump-nesting and is a habit of several species of birds, such as the moorhen. It has been discovered recently that American magpies and pinyon jays actually carry some of their eggs from their own nests and drop them into their neighbours' nests. This could be a way of putting one's eggs in more than one basket, in case one's own basket is destroyed.

Some birds behave like the cuckoo, using the nests of other species, but only on an occasional basis. This type of brood parasitism can only occur when the requirements of the infant birds are similar. A 'cuckoo in the nest' is an extra burden for the 'host'. It may mean no more than an extra mouth to feed, but some cuckoos either remove a host egg as they lay their own, or the hatchling parasite kills or ejects its nest mates to ensure that its foster parents will rear it successfully. To counter these threats, the host attempts to foil the parasite by building inaccessible or impenetrable nests or attacking strange birds that come near their nests. The brood parasites overcome these defences by stealth and agility. The females sneak their eggs into the nest when the owners are absent, or wait for the males to lure the hosts away, giving them time to lay – a process that takes only a few seconds to complete. And the European cuckoo and American cowbirds are not put off by the sight of a domed nest; they simply lay their eggs through the entrance.

When a bird discovers a parasite egg, it may abandon the nest, eject the egg or build a false floor over it. However, it can be fooled into accepting it if the parasite egg resembles its own eggs – sometimes it is an almost perfect match. This is the most interesting and striking aspect of the cuckoo habit and admiration for its perfection helps to offset distaste for what is often seen as 'cruel' behaviour.

Most cuckoo species have a range of egg colours but each individual lays eggs to match one particular host. It is believed that the female inherits egg colour from her mother and chooses the right nest in which to lay by searching for the species that reared her. The large hawk-cuckoo of Asia has two egg colours: brown eggs to match those of shortwings and spiderhunters and blue eggs for laughing thrushes. Even where the hawk-cuckoos and the two groups of hosts breed in the same locality each female selects the appropriate host for her eggs.

The situation is more complicated for the European cuckoo, which parasitizes far too many species to be able to match all their egg colours. It has two main colour types: one is spotted with variations to match the eggs of pipits, wagtails and robins; the other is plain blue to match those of redstarts, bramblings, great reed warblers and dunnocks.

In each area there are a few favourite hosts which the local cuckoos match. In eastern Europe these are the brambling, redstart and great reed warbler and

A moorhen may lay some of her eggs in the nests of other moorhens. This is called dump-nesting and it may be an insurance against her own nest being destroyed.

The European cuckoo is unpopular because of its habit of laying its eggs in the nests of other birds – such as the dunnock nest shown here – but this is a marvellous example of biological adaptation. The cuckoo hatches first and throws the other nestlings and eggs out of the nest, so that it receives the undivided attention of its foster parents. Its large orange-lined mouth acts as a strong stimulus and even birds which are not its fosterers are lured into giving it food.

there are subsidiary hosts which tolerate eggs 'designed' for the main hosts. For example, the blue 'redstart' egg is accepted by wheatears and pied flycatchers. Surprisingly, western European cuckoos do not have a blue egg to match those of the dunnock although it is one of their main hosts. Parasitism in such cases is successful because the dunnock is very tolerant of mismatched eggs.

Apart from colour matching, cuckoo eggs are specially adapted by having an extra strong shell. The eggs of the European cuckoo and the crested cuckoos of Africa and India have shells twice as thick as those of their hosts and they are given added strength by the nearly spherical shape. This is an adaptation to the way that the eggs are dropped into the nest often from a height of several centimetres.

The laying of a cuckoo egg is often held up for a day, so the embryo has already started to develop. Added to this, incubation is usually one or two days shorter than for the host eggs, so the cuckoo nestling hatches first and has a head start on its rivals.

The cuckoo habit has also developed among the nesting insects, especially the social insects. The

cuckoo wasp is a solitary wasp which enters the nest of one of the social wasps to lay its eggs in the brood cells. When the larvae hatch, they kill the rightful occupants of the cells and are reared in their place by the worker wasps. Sometimes a bumblebee queen searching for a suitable place to build her nest will find the newly established nest of another queen; she kills her and takes over the nest. This is regular behaviour for *Bombus hyperboreus* of the Arctic, who takes over the nests of *Bombus polaris*. She does not produce any workers of her own, so all her eggs become males and females which are reared by the *polaris* workers.

Cuckoo bees of the genus *Psithyrus* are almost identical to 'normal' bumblebees, but close inspection shows that they do not have pollen-baskets on the hind legs. Neither can they manufacture wax for constructing cells in the comb. The queen cuckoo bee seeks a bumblebee nest in spring, before there are many workers to defend it, and sneaks in. Although stronger than the workers and having thick plates on her body armour, she could be overpowered and thrown out or killed if discovered, so she feigns death if approached. Once inside the nest, she hides under the comb until she has acquired the identifying smell of the colony. Then she destroys any *Bombus* eggs she can find and uses wax from the host comb to make her own cells in which she lays her eggs. She may have to defend them against the host workers, but otherwise it is in her interest to leave the workers alone as they will be responsible for rearing her offspring, but she eventually slays the host queen and becomes the only egg-layer.

In the ant world, there are 'slavemaking' species which take over other species' nests and recruit the workers for their own ends. After the mating flight, the queen slavemaker or robber ant behaves like the cuckoo bees. She seeks out a nest of the negro ant and steals inside. There, she appropriates some pupae and the workers which emerge from them become her employees. The negro ant queen is soon killed but the supply of slaves is maintained by bands of slavemaker workers sallying out of the nest and raiding neighbouring negro ant nests to steal their pupae.

Slavemaker ant colonies can survive without their slaves, but other parasite ants have dispensed completely with their own workers. The queens rely on the host workers to feed her and to care for the eggs and young. The host queen soon dies, probably because she is starved or killed by her own workers, although it is a mystery how they are turned against her. The result is that the colony soon dies out, but not before the parasite's eggs have hatched and the young ants have left the nest to mate and invade new colonies.

CHAPTER 6
How many eggs?

The record for egg-laying in domestic chickens is held by a Black Orpington hen that laid 361 eggs in 364 days. This amazing productivity was the result of centuries of selective breeding and intensive feeding, since the jungle fowl, the wild progenitor of the domestic chicken that lives in the jungles of Asia, lays fewer than half a dozen eggs in its annual clutch. An animal's egg-laying capacity is called its fecundity. Compared with the fecundity of many invertebrates, even the record-breaking Black Orpington pales into insignificance: the American oyster sheds five hundred million eggs in a year and the common mussel lays 12 million in a quarter of an hour.

Certain parasitic animals are even more prolific. The tapeworms that live in the intestines of humans and other species consist of a scolex, or head, armed with hooks and suckers that stick to the wall of the intestine, and a flattened body made up of short segments called proglottids. Each proglottis contains male and female reproductive organs and eventually ripens into a bag of eggs. The fish tapeworm can grow to 20 metres, longer than the intestine of its host in which it lies coiled. The proglottids break off as they ripen and pass out of the intestine. The process is continuous and it has been calculated that a ten-year-old fish tapeworm may have formed as much as seven kilometres of proglottids, containing a prodigious two thousand million eggs.

At the other end of the scale, there are animals that lay very few eggs during their lifespan. For example, the sheep ked, a wingless fly that sucks the blood of sheep, produces only ten or 12 eggs in a lifetime. Within this range of egg-laying there are animals laying eggs in varying numbers, in large or small batches and at different frequencies. Among birds, comparison can be made between the giant albatross which lays a single egg in alternate years, and the European blackbird which produces two, three or

Right: a clutch of 12 blue tit eggs in a nestbox. This is an average number for the species, but up to 19 may be laid by one bird. The largest clutches are laid early in the breeding season to coincide with peak food availability for the young.

Below right: the clutch of a wood pigeon numbers only one to three eggs. Each female lays up to three clutches a year, but many are lost to predators.

Below left: a male stickleback bloated with the larva of a tapeworm. The larva will develop into an adult only if the stickleback is eaten by a bird. Because the chances of completing this complicated life cycle are so slender, tapeworms produce an enormous number of eggs.

sometimes four clutches of four to five eggs each year, or the snowy owl which may lay up to 14 eggs in a single annual clutch.

The species' fecundity also depends, of course, on the number of breeding seasons that the female lives through; thus the slow laying rate of the albatross is balanced by its long life. All in all, the size of a clutch of eggs and the number laid in a female's lifetime is adjusted through evolution so that she contributes as many offspring to the next generation as possible. There are two factors at work. Within a species, the females need to produce enough offspring to maintain the population. Also, the number of eggs laid will vary according to the climate, availability of food and so on.

Taking the first factor, the fecundity of a species is the result of natural selection which has led to a rate of egg production that will ensure enough offspring to continue the animal's lineage. The rate of egg production will depend partly on the lifestyle of the species. In a stable population, a pair of animals has to be replaced by two offspring surviving to adulthood. Consider a starfish shedding forty million eggs into the sea each year: if the population is to remain stable,

Left: two clusters of eggs laid by woundwort shieldbugs, *Eysarcoris fabricii*, on the underside of a leaf. These are small clutches for an insect.

Below left: part of a large clutch of moth eggs. Although nearly all these eggs will hatch, many of the caterpillars will be eaten before they are fully grown.

Right: *Cyclops*, the one-eyed freshwater crustacean, carries its eggs in twin pods. It lays fewer eggs than its relatives that shed their eggs into the water.

99.999999 per cent must die or the sea bed will be overrun with starfish. Since starfish eggs and larvae are surrounded by predators as they float in the sea, there would seem to be some relationship between mortality and fecundity, in that the more dangerous the life the more eggs must be laid.

The fecundity of the various species of ichneumon fly which parasitize the Swaine jack pine sawfly, *Neodiprion swainei*, ranges from 30 to one thousand eggs laid in a lifetime. Those flies that lay their eggs in young sawfly larvae have a high fecundity which is set against high losses due to the host larvae themselves having a high mortality rate through predation. Flies which attack older sawfly larvae, after they have dropped from the jack pine leaves to the forest floor, lay fewer eggs. These older sawfly larvae are well concealed and suffer less mortality; the flies that parasitize them lay fewer eggs since a greater proportion will reach adulthood.

Ecologists have found, however, that high fecundity is not just an evolutionary response to the eggs and larvae being defenceless and falling victim to many kinds of predators and natural disasters – in fact it is the other way round. An animal has a finite amount of energy and materials that it can deploy for manufacturing eggs, so if it reduces the size of each egg it will be able to lay more of them. This will give the animal an advantage (in the evolutionary sense of

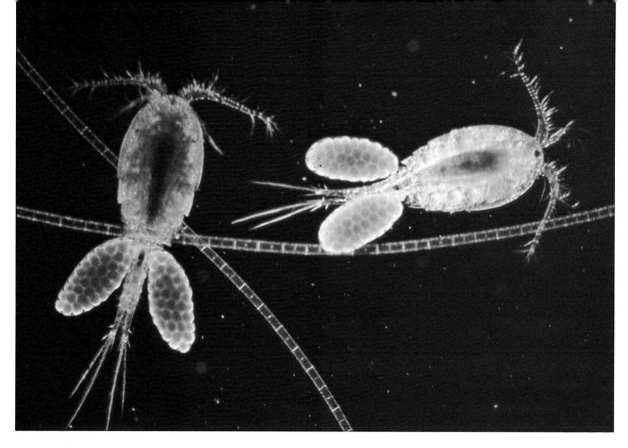

leaving more offspring) over its fellows who are laying fewer eggs. Ultimately, of course, there comes a point when the egg becomes too small to have much chance of surviving. As the ecologist Paul Colinvaux has pointed out in his book of essays, *Why Big Fierce Animals Are Rare*, 'a high death rate for the tiny, helpless young is a consequence of the thousands of tiny eggs, not a cause'. This is because the creatures that hatch from small eggs are relatively feeble and easily eaten or destroyed by natural hazards.

One advantage of laying large numbers of small eggs is that the species has the ability to survive in an unpredictable or harsh environment and to colonize new habitats. A catastrophic drought or change in saltiness of the sea can wipe out most of a population, so explosive bouts of egg-laying are needed to rebuild the population. Marine animals often lay small eggs that float in the sea and quickly hatch into free-swimming larvae. Both eggs and larvae are easily dispersed into new habitats. Species such as barnacles and mussels are therefore good colonizers and are the first to settle on a newly built oil rig or to reappear on a shoreline after devastating pollution.

An alternative option is to lay fewer but larger eggs, each containing more nourishment, so that the hatchlings are launched with a better chance of survival. Also, if fewer eggs are laid, part of the adult's energy reserves can be reallocated to caring for its

offspring by depositing them in a nest or some other protected place and defending them against danger.

Parasites and polar invertebrates are good examples of the extremes of these habits. Parasites typically lay large numbers of eggs because of the uncertainty of their lifestyles; they often need to move between two or more hosts to complete their life cycles. For instance, the fish tapeworm's eggs are shed from the digestive tract of the mammal host and, if they fall into water, they may be eaten by the common freshwater crustacean *Cyclops*. The chances of being laid in water and then being eaten by the right animal are, however, very slender. The crustacean must then be eaten by a suitable fish, such as a pike. The immature tapeworm bores through the pike's intestine and lodges in the muscles, where it stays until the fish is eaten by a human or some other mammal and the tapeworm matures. At each stage the odds against the tapeworm making the transition from one host to another are very high, hence the need for a huge production of eggs to ensure the survival of a few tapeworms to adulthood.

The general rule that the majority of marine invertebrates shed their eggs into the sea is broken in the cold waters of the polar seas. In the Arctic, most invertebrates lay large yolky eggs which lie on the sea bed and the embryos complete their development before hatching. In the Antarctic, invertebrates often

brood their eggs until they hatch in the adult form, or give birth to live young. These strategies reduce the mortality of the eggs and young, so polar species lay fewer eggs than their temperate and tropical relatives. The penalty is a reduced ability to colonize new habitats, but this form of development is well suited to the polar environment.

Polar invertebrate eggs are usually much bigger than their temperate equivalents. For example, those of the Antarctic shrimp *Chorismus antarcticus* are seven times larger than those of the temperate water 'pink shrimp' *Pandalus*. Larger eggs take longer to develop than small ones. This is advantageous for the Antarctic shrimp as at high latitudes there is only a short summer period of abundance and floating larvae would be unable to complete their development before the winter. The slow development of large yolky eggs means that, even though they are laid in summer, the juvenile animals will not emerge until the following spring when the sea is becoming productive again.

Theories relating a species' lifestyle to the number of eggs it lays have to be treated with some caution, however, because of the variation between the number of eggs laid by the individuals within a single species. For instance, recently a biologist found that, for one species of limpet, there was more variation in clutch size between populations living on different parts of the shore than there were between four

species of limpet whose egg production had previously been considered to be related to their lifestyles. Among crustaceans such as crabs, shrimps and prawns, egg production is also related to the size of the female's body, independently of any other factor; larger and therefore older females lay more eggs than smaller individuals, simply because they have more room in their bodies for eggs.

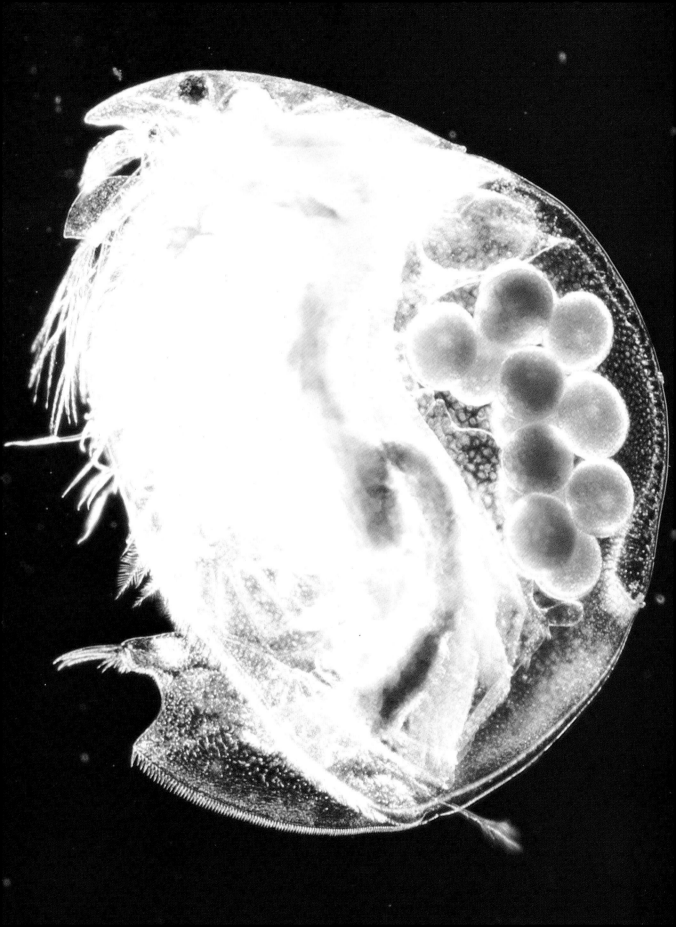

For other species slow development can be a disadvantage because of the mortality suffered by the eggs. On the other hand, small eggs tend to hatch at an early stage of development and the tiny young are also vulnerable. Some molluscs have solved this problem by laying small eggs that hatch out advanced larvae. This seemingly paradoxical strategy has been achieved by providing the developing embryos with food from outside the egg. Among the rockshells – marine molluscs that include the dogwhelks – up to 50 per cent of the eggs in one capsule may die. Instead of being a total loss of reproductive effort and a source of infection, the dead embryos are consumed by the living ones. They continue to develop quickly by virtue of their small size, but they hatch out in an advanced state of development because they have received extra food. One of the murex molluscs produces either swimming larvae or crawling young depending on whether or not the embryos in an egg capsule have fed on dead embryos. Other species regularly produce a few eggs that die and act as 'nurse eggs' for the survivors, and one of the sea slugs provides its egg mass with a ribbon of yolk. In the summer, when food in the sea is plentiful, the parent sea slug saves energy by providing a small yolk ribbon and the eggs hatch into swimming larvae, but in autumn when food is becoming scarce, it provides more yolk and the eggs hatch out as crawling young which are tiny replicas of the adults.

An individual's egg-laying capacity is further modified by environmental factors. The starfish *Leptasterias hexactis* broods its eggs until they are ready to hatch. It carries them under the centre of its body which is raised off the rock, and the more eggs the starfish carries the fewer tube feet can be used for clinging to the rock. In this vulnerable state it runs the risk of being swept away by the waves, thereby losing its eggs and perhaps its own life. Consequently, starfish living on exposed, wave-swept shores brood smaller numbers of eggs and cling more tightly than those on sheltered shores.

Temperature, light and humidity can also affect egg-laying. In recent years a new household pest has appeared in the form of a tropical relative of the booklouse. It originally came from Africa but is now cosmopolitan and lives in kitchens where it thrives on flour, breakfast cereals, sugar and other stored foods. Modern kitchens, which no longer have cool larders, are usually well insulated and poorly ventilated, so the atmosphere is warm and damp. (Brand-new houses are best for the booklouse, because the plasterwork is still damp.) At temperatures of 25–30 degrees centigrade and a relative humidity of 75 per cent, a female booklouse lays about 110 eggs in her six months of life. At her peak she is laying two eggs a

day, each one a third as long as her body. But, as temperature and humidity fall, egg production drops and at temperatures below 20 degrees centigrade it virtually ceases.

The state of nutrition of the female is, however, the most important factor that affects the number of eggs laid. This is self-evident: if she does not have enough food reserves she will not be able to produce eggs, as any poultry-keeper will know. We have already seen how the male can significantly help egg production by feeding his mate, sometimes inadvertently and suicidally in the case of spiders and the praying mantis, but these are the exceptions. The female must usually rely on her own resources and reproduction cannot proceed unless she is well fed.

Some insects lay down food reserves during the larval stage, so the adult female does not need to feed. Visitors to the Arctic must wonder how such a desolate place supports such an enormous population of biting insects; the arrival of humans is the signal for swarms of mosquitoes and blackflies to take the blood meal that will nourish their developing eggs. The surprise is that such large numbers of these insects can be maintained where there are so few native warm-blooded animals. The answer is that they use the protein laid down as body reserves when they were aquatic larvae to lay a small number of eggs. This ensures the survival of the species until a suitable blood donor appears and enables the insects to lay extra batches of eggs.

Birds: a special case

Counting the eggs laid by an invertebrate is no easy matter, especially when millions are shed into the sea, but the number of eggs in a bird's nest must be one of the easiest kinds of biological data to collect. Every bird book gives figures for clutch sizes and more detailed records yield information about the number of eggs laid according to the age and experience of the birds, food supply, climate and other factors. It is also relatively easy to record the number of young birds that fledge from a clutch, or those that survive the tricky first year of life, by analyzing recoveries of bird rings. By using such information it has been possible to investigate the evolution of egg production in birds.

For each species, egg production or clutch size will have evolved so that it yields the greatest number of offspring to contribute to the next generation. Even if wild birds could produce masses of eggs, like the record-breaking Black Orpington, this would not, by itself, lead to flocks of young birds swarming the countryside. There is a limit to the amount of food that parent birds can collect for their nestlings, so an

Many birds lay a fixed number of eggs in a clutch, although in adverse circumstances they may lay fewer than normal. Young birds may also lay smaller clutches. The snow petrel of the Antarctic (*above left*), like all members of the petrel and albatross order, lays a single egg and the red-throated diver (*above right*) lays two eggs. The lesser black-backed gull (*below left*) lays three eggs. Like many waders, Sanderlings (*below right*) lay clutches of four.

extra large clutch may be a positive disadvantage because the young will be undernourished and may even starve. The result may be fewer young being raised than from a smaller clutch.

The relationship between clutch size and the number of young that survive is nicely illustrated by swifts nesting in Britain. The clutch consists of two or three eggs and in fine summers, when the flying insects that form the swifts' food are abundant, almost all young are raised successfully. Hence pairs that lay three eggs have the advantage. In cold, wet summers, the parents bring fewer meals per day and the nestling swifts weaken and die. The chances of being undernourished are greater when there are more mouths to fill, so pairs laying two eggs raise more young than those laying three. If British weather were to become more dependable, the swifts would evolve a fixed clutch size: two if it was dependably bad and three if it was dependably good. In continental Europe, where the weather is better, swifts more commonly lay three-egg clutches and may lay four eggs – a rare occurrence in Britain.

There is also the possibility that the parents of a large clutch may expend so much energy trying to raise the brood that subsequent breeding is impaired. This is suggested by an experiment with rooks, which normally raise two fledglings: when pairs were given extra nestlings to rear, their breeding was less successful in the following year. Birds who raise large broods may even put their own lives at risk, by becoming so run down that they do not survive the next winter. Again, researchers discovered that house martins that had raised two broods were less likely to return from winter quarters, where presumably they had died, than those that had reared just one brood.

Clutch size can vary within a species and an individual bird may lay different-sized clutches from one year to the next. There is little indication that food supply determines the number of eggs laid on an individual basis; if extra food is provided, the bird advances the egg-laying date rather than increasing the size of its clutch. One exception is the snowy owl, which depends largely on lemmings during the breeding season. When lemmings are abundant, the owls lay large clutches, but when the lemming populations crash, the owls may not lay any eggs.

Adapting the clutch size to local food supply is not the same as the evolution of the clutch size as part of the bird's inherited adaptation to its environment. A number of birds have a fixed clutch size: the petrel family and large eagles lay only one egg; some penguins and gulls lay two eggs, while related species lay three eggs and plovers lay four. Sometimes full adaptation has not been achieved and within the gannet family of sea birds clutch size is not always appropriate to the ability to raise young. Boobies are tropical gannets and tropical waters are usually less productive than temperate seas. Masked and brown boobies lay and hatch two eggs, but they almost always lose the younger nestling which is killed by its elder sibling because the parents cannot provide enough food to satisfy them both. The red-footed booby has adapted more thoroughly and lays only a single egg. In contrast, the Peruvian booby lays four eggs and rears its young on fish caught in the cool, richly productive waters of the Humboldt Current that sweeps up the Pacific coast of South America. This is the current that sustains the enormous shoals of anchovetas off the coast of Peru and periodically

devastates them when it fails and is replaced by a flow of warm water, in the phenomenon of *el Nino*. When this happens, the Peruvian boobies lose all their offspring and may lose their own lives too. The temperate water gannets also appear not to be fully aligned with their environment. They lay only a single egg, although they have more food available. This may be because they put greater effort into feeding the single chick and providing it with a good layer of fat to nourish it when it leaves the nest, but Atlantic gannets have proved capable of rearing two nestlings when given an extra egg.

There is a tendency for land birds living in northern latitudes to lay more eggs than those living nearer the equator; this holds both for closely related species and individuals of species that are found across a wide range of latitudes. Longer summer days at higher latitudes are one factor. The more hours there are of daylight, the more time the parents have for gathering food and the more mouths they can feed. Birds have, therefore, evolved clutch sizes that are appropriate to their food-gathering time. In some species there is a distinct correlation between clutch size and latitude. Robins living in the Canary Islands have an average clutch of 3.5; those in Spain lay 4.9 eggs, in the Netherlands 5.9 eggs, and in Finland 6.3 eggs.

Day-length cannot be the whole explanation because the tendency also holds for owls which have shorter nights to hunt in! The alternative explanation seems to lie in the equable nature of tropical habitats when compared with northern latitudes, where a harsh winter is followed by a spring flush of food. In the north, many birds die in the winter so there is less competition for the abundant food in the breeding season and the survivors can raise larger broods. There is a similar difference within the tropics where birds nesting in savannahs, which have marked seasons of drought followed by rains, lay larger clutches than closely related species nesting in the even climate of rain forests.

Birds that can vary their clutch size are called indeterminate layers. If eggs are taken away as they are laid, they are replaced. By such means, a yellow-shafted flicker (an American woodpecker) was persuaded to lay 71 eggs in 73 days, instead of the usual five to ten eggs. Others have a limited capacity to replace eggs. Penguins lay an extra egg if one is removed and gulls can lay up to three replacement eggs. Other birds are determinate layers: they lay a full clutch and no more. They can, however, lay a repeat clutch if the first is lost, although fewer eggs may be laid. Courtship starts again and a new nest may be built. Exceptions include the albatrosses and petrels and the large vultures who will not attempt to lay again until the next season.

Enemies of eggs

An egg is very vulnerable. Immobile, defenceless and rich in nourishment – as Shakespeare rightly said, it is 'full of meat' – it makes an excellent meal. We have already seen that the losses of eggs can be enormous, even among species that protect their clutches. Tropical birds, for instance, lose as many as half of their eggs. Small eggs face the danger of being accidentally eaten by all manner of animals, including vegetarians, such as the winkles grazing on dogwhelk eggs mentioned on page 26. Similarly, the eggs of butterflies, moths and bugs which have been laid on leaves may be ingested by caterpillars, but not only accidentally – some caterpillars are predators and regularly eat the eggs of other insects.

Many insect eggs fail because, in spite of their size, they are often invaded by parasites. The habits of the chalcid wasp *Trichogramma* have already been described, but stranger still is the life history of the mite *Adactylidium*. The female mites attach themselves to the eggs of thrips, minute insects which

Above: an egg of a great grey slug with three larvae of a parasitic fly growing inside.

Left: instead of hatching out caterpillars, these drinker moth eggs have produced tiny parasitic *Telenomus* wasps.

Below: a brimstone butterfly egg is menaced by a parasite egg laid beside it. The parasite larva will hatch out first and feed on the larger egg.

91

can be serious plant pests and are well known as 'thunderflies' that land on our skin and tickle us in hot weather. A single thrips egg provides each mite with all the food she will consume in her lifetime, during which she produces a brood of five to eight daughters and one son. These offspring develop in their mother's body and devour her tissues. When mature, they emerge. The son immediately dies because his life's work is done; he has already impregnated his sisters, who disperse in search of more thrips eggs to repeat the cycle.

It is not at all easy to list the predators of the eggs of insects and other invertebrates, since they fall prey to any animal that accepts morsels of food in the size range of the eggs, while eggs floating in the sea may be snapped up by any fish or carnivorous invertebrate living in the open water. One way of discovering the predators of a species' eggs is to mark several eggs with a radioactive substance and then investigate which predators become radioactive. In one experiment, thousands of mosquito eggs were labelled with radioactive phosphorus and set out in a mosquito breeding area. A later check showed that the radioactivity had been transferred to two species of ant and four species of ground beetle.

Ants are probably the best-known predators of insect eggs. A single tree may be visited by thousands of ants in a day and they search every crevice where eggs may have been deposited. But ants also protect eggs against predators. It is well known that some ants defend aphids, as well as other plant-sucking bugs, in return for their honeydew, and they carry aphid eggs into the safety of their nests, guarding them through the winter until they hatch. Because ants are so vigilant they sometimes interfere with control programmes that use parasites and predators against agricultural pests. In Kenya, scale insects attacking coffee plants were brought under control by ladybirds, but only when sticky bands were wrapped around the coffee stems to prevent ants from climbing up and defending the scale insects.

The eggs of vertebrates are similarly attacked by a variety of predators; some merely take advantage of a good meal when they find a nest, but others are specialist egg-eaters. Hedgehogs, which normally feed on insects, rob the nests of gulls and game birds. The

Above left: Kittlitz' sand plover lays its clutch on the bare ground. When a predator approaches, it conceals the eggs by kicking sand and pebbles over them.

Above right: the egg-eating snake is a specialist egg predator that engulfs eggs wider than its own body, crushes the shells and swallows the contents.

Right: a cat-eyed snake of the American rain forests eats leaf frog eggs despite their jelly covering.

Left: the hedgehog is only a casual predator of birds' eggs, but if it finds a nest during its nightly search for insects, the eggs will provide an easy and nutritious meal.

Galapagos mockingbird pecks open the eggs of the waved albatross, which are particularly vulnerable because they are laid in nests on the ground. A clutch of eggs is worth making an effort to get. The Nile crocodile buries its eggs and defends them, but the nest may still be robbed by monitor lizards, baboons, honey badgers and mongooses. The Roman author Diodorus believed that the Egyptian mongoose helped to get rid of crocodiles solely out of a desire to assist the human race. He wrote: 'Were it not for the service it thus renders to the country, the river would become unapproachable from the multitude of crocodiles.'

The shells of birds' eggs cannot be too thick or the chicks will not be able to hatch out. They are strong enough to avoid breakage when shuffled in the nest, but they cannot resist attack. Hedgehogs and long-jawed carnivores such as foxes can get their teeth around a bird's egg and crush it, but snub-nosed mongooses have difficulty with all but the smallest eggs. Several species of mongooses have overcome this problem by developing methods of smashing eggs. The marsh mongoose grabs the egg in its front paws, rears up and hurls it at the ground. Alternatively, it and other species of mongoose throw eggs backwards between their hind legs. Skunks perform in the same way. Tame banded mongooses were given ostrich eggs by the zoologist Jane Goodall; having failed, not surprisingly, to throw them between their legs, they threw stones at them. Jane Goodall's interest in how mongooses dealt with ostrich eggs stemmed from her observations of Egyptian vultures. Like the mongooses, the vultures' usual technique for smashing eggs is to throw them at the ground, but they throw stones at ostrich eggs.

Some sea snakes live entirely on fish eggs but the most specialized egg-eater is the egg-eating snake of Africa and India. Its neck is the size of a man's finger, but it can swallow whole a hen's egg with a diameter of 45 millimetres because a characteristic of snakes is that they can unhinge their jaws to swallow objects wider than themselves. When a snake finds an egg, it smells it to make sure that it is not addled, coils its body around it to hold it steady and then engulfs it with slow gulps. Once swallowed, the egg is opened by a remarkable arrangement of projections from the

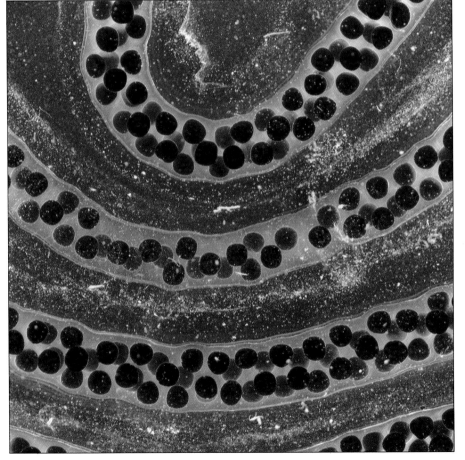

Left: the jelly surrounding the eggs of amphibians gives some protection against predators, but certain toad eggs also contain a poison. A few of the eggs will be sacrificed in this way to deter a predator from eating the rest.

Right: a group of cat fleas with one egg. Eggs can transmit disease as well as being poisonous. Flea eggs, for example, can harbour disease organisms such as those of plague; they are transmitted from one generation of fleas to the next and thence to a new host.

Below: poisonous eggs are characteristic of species in which the adult is also toxic, such as this burnet moth. It seems that the developing embryos are, like the adults, immune to their own poisons.

snake's backbone that protrude like teeth through the roof of its throat. The first 17 or 18 of these 'teeth' are long and sharp. The egg is held against them by the throat muscles and the snake bends it head to and fro to saw through the eggshell and release the contents. The empty shell is then moved to the back 'teeth', which are blunt, and they compress the shell into a rolled-up sausage so that it can be easily ejected.

With all these enemies it is easy to see why animals make such strenuous efforts to protect their eggs. No defence is perfect: many invertebrates and fishes and most reptiles bury their eggs or hide them in nooks and crannies, but all sorts of animals seek them out. Active defence is not always successful because the parent itself may be caught and eaten. However, the Australian blue-ringed octopus is not only capable of delivering a lethal bite by injecting a fast-acting neurotoxin, but also adds this poison to its eggs. Surprisingly, very few other animals use this method of defending their eggs against enemies. Those that do include the garden tiger moth, several toads, the pufferfish and the starfish *Bougainvillia*. Like the blue-ringed octopus, all of these animals use poison to defend themselves as adults and their immunity to their own poisons is presumably present in the developing embryos. Other animals may have attempted to evolve poisonous eggs but never found a way of doing it without harming the embryos. One that has found an alternative solution is a species of bug which plasters its eggs with camphor collected from camphor plants – to act as an insect repellent.

CHAPTER 7

The marvel of development

Since the seventeenth century, when William Harvey established that the egg was the starting point for the development of every new animal and Anton van Leeuwenhoek argued that development was triggered by fertilization with a sperm, embryologists studying the development of eggs have been filled with a sense of wonder at the beauty and mystery of the process. The sense of wonder was evoked by observations through the microscope of the way that the simple, single-celled egg develops by stages into a perfect, living animal. The mystery was how this transformation takes place. How does a fertilized egg turn into an elephant so totally different in size and form? By what mechanism is it determined that there will be a trunk at one end and a tail at the other? The embryo must 'know' how to construct an eye or an ear or where to grow its legs and how to connect up the nerves and blood vessels.

The study of embryos is a means of shedding light on the problem of how cells grow, develop and differentiate into tissues. It has been given a great fillip by the need to investigate defects caused by embryonic malformation, such as spina bifida, and the possibility of inducing the body to regenerate damaged organs. But the central problem for embryologists has been how the egg, this unique cell, is transformed into an animal – a complex being of millions of cells, that moves, breathes and interacts with its environment – using information stored within itself.

The early microscopes revealed the nature of the egg and sperm, but the simple research techniques of the time could not help to explain the nature of embryonic development and this continued to be the subject of highly speculative theorizing. There were several schools of thought. The preformationists held that an animal developed from a fully formed but microscopic organism – known as the homunculus in humans – with all its organs in place. Fertilization provided the stimulus for it to grow into an adult. The spermists believed that the homunculus (or its equivalent in other animals) was carried in the sperm and the ovists that it was carried in the egg.

The epigenesists, on the other hand, held that there

was no preformed structure but that the developing embryo became progressively more complex and organized. The epigenesists had the right idea: a single egg cell is transformed into a complex organism by the fashioning of specialized tissues which become limbs, internal organs and so on. But the preformationists were not wholly wrong either and their thesis held the germ of modern genetic theory: information for organizing the embryo is carried from one generation to the next, not as a homunculus but as blueprints in the chromosomes.

As microscope techniques improved, the story of embryology was slowly revealed, but most texts on the subject are extremely difficult to understand. For example, in the early stages of development especially, structures which have no equivalent in the adult animal appear and disappear. They are therefore unfamiliar and their names mean nothing to anyone but an embryologist. It is also inappropriate to call a lump of tissue a brain or a heart, even though that is what it will eventually become, so embryologists talk of the presumptive brain or heart. None of this makes for easy reading and although the gradual shaping of an animal is one of the marvels of the natural world, embryology stays in the remote atmosphere of the laboratory where it remains a Cinderella subject passed over by most natural history books. Yet it has much to offer: everyone has at some time wondered 'where they came from'.

Seeking the answer to this question is fraught with technical and ethical problems. Luckily, development follows the same basic steps in all animals, so it is possible to learn much about human embryology by watching frogspawn turn into tadpoles or a hen developing inside an egg. Hippocrates, the Greek physician universally regarded as the Father of Medicine, realized this. He wrote: 'Take 20 or more eggs and let them be incubated by two or more hens. Then each day from the second to that of hatching, remove an egg, break it, and examine it. You will find exactly as I say, for the nature of the bird can be likened to that of a man.'

Around two thousand years later, when the theory of evolution was introduced, zoologists came to

The developing eggs of a three-spined stickleback at four days (*above*), six days (*centre*) and seven days (*below*) after fertilization. Stickleback eggs hatch at 5–12 days, depending on the temperature. Droplets of nourishing fat can be clearly seen, as can the embryos' large eyes. At first the eyes are transparent cups, but they become black as pigment is laid down.

believe that the development of the embryo repeated the course of the animal's evolution. For instance, the embryos of humans and all other vertebrates possess at a particular stage the rudiments of gills and a fish-like tail which was supposed to indicate that they had a common fishy ancestor. Embryology could, therefore, be combined with the fossil record to produce a 'tree of life' showing the relationship between different groups of animals.

It was to shed light on the origin of birds that Wilson, Bowers and Cherry-Gerrard, of Scott's last expedition, set out on the 'worst journey in the world' to collect emperor penguin eggs in the awful conditions of an Antarctic winter. Emperor penguins were thought to be primitive birds, so their embryos would, hopefully, have characteristics of the ancestors of modern birds. Unfortunately for the theory, penguins are not primitive. On the contrary, they are very specialized birds, and the embryos could not provide an answer to the problem of the birds' ancestry.

It is now known that the theory that sparked off this hazardous journey was incorrect; the sequence of embryonic development does not follow evolutionary history. The gills and tail in the human embryo are not signs of a fishy ancestry. There is little resemblance to an adult fish; rather, the human embryo resembles a fish embryo, which shows that development has been very conservative and has hardly altered despite great evolutionary changes in adult animals. Development follows a common plan: every species starts as an egg and the early embryos resemble each other closely; thereafter the similarity weakens depending on the final resemblance of the adults. A vertebrate and a mollusc go their separate ways during cleavage (see page 99), while reptile and bird embryos are virtually indistinguishable until well into development and identification of species is not always easy even after hatching.

The simplest way to describe how an animal takes shape in an egg is to start with the frog's egg. Frogspawn has been used for teaching the rudiments of embryology for many years and is still collected by children for the pleasure of watching it change first into tadpoles and then into froglets. Because the eggs are large and the jelly covering transparent, observation and manipulation are much easier than with the shell-covered eggs of birds. However, a tadpole hatches out when development is incomplete – it is only halfway to becoming a frog – so to understand the later stages it is necessary to switch to the hen's egg, where the embryo remains inside the shell and a number of extra structures provide it with a life-support system until development is complete and it can safely emerge.

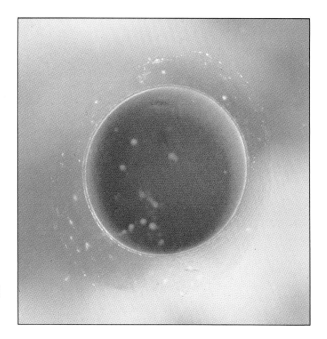

The early development of the European common frog.
Above: a new-laid egg surrounded by jelly; it has been fertilized, but cell division has not yet begun.
Above centre: the two-cell stage after the first cell division has taken place.
Above right: a second vertical cell division results in a four-cell embryo.
Right: the next cell division is horizontal and cuts above the middle of the embryo. The larger cells will become the yolk.
Far right: repeated cell divisions create a ball of small cells called the blastula.

The sequence of development can be followed most easily if the general outline of events is first understood. The major steps following fertilization are *cleavage*, a period of rapid cell division without growth that creates a hollow ball called a blastula; *gastrulation*, the tucking and folding of the blastula to make a two or three-layered structure called the gastrula; and *organogenesis*, the formation of organs and growth of the embryo that takes up the greater part of its life. Three processes are involved: *cell division* to increase the bulk of the embryo as it grows, *differentiation* of these cells into the many kinds of tissues and finally *morphogenesis* (a word derived from the Greek words for 'shape' or 'form' and 'production') as the tissues are moulded into the organs that make up the complete animal.

The general plan of the embryo is decided even before it starts to develop. The point where the sperm entered the egg is marked by a tiny spot, the sperm aster, whose position will determine the orientation of the first division of the fertilized egg or zygote and its subsequent development into an embryo. About

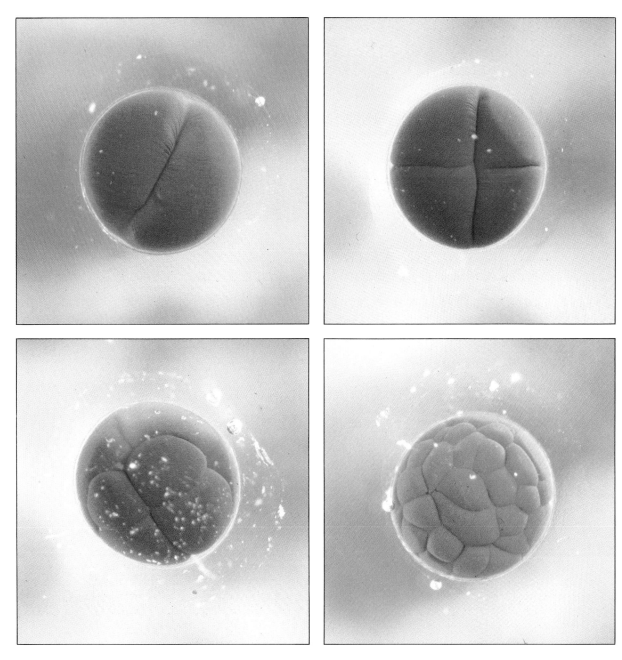

30 minutes after fertilization, pigment on the zygote's surface moves towards the sperm aster to expose an area of cytoplasm called the grey crescent on the opposite side of the egg. The grey crescent and other regions in the zygote's cytoplasm are destined to become organs in the later embryo: some tissues formed from the grey crescent will become the notochord – a rod in the back of the embryo that keeps it rigid until it is replaced by the backbone. There are also changes taking place inside the zygote. The nucleus comes to lie in one half, the upper, black part of the frog egg. This 'animal hemisphere' is where the embryo will start to develop and it contrasts with the pale grey yolky part which will become the large mass of yolk tissue that nourishes the embryo as it grows.

During the next stage of development the single, outsize cell of the zygote turns into a mass of tissue with normal-sized cells. This is cleavage, a series of cell divisions that continues apace, with the egg maintaining the same size while individual cells get progressively smaller.

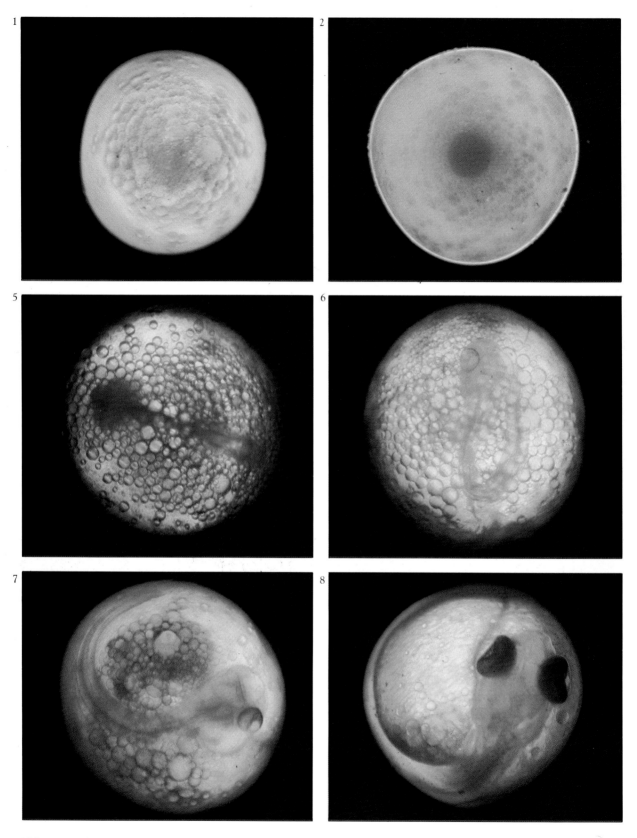

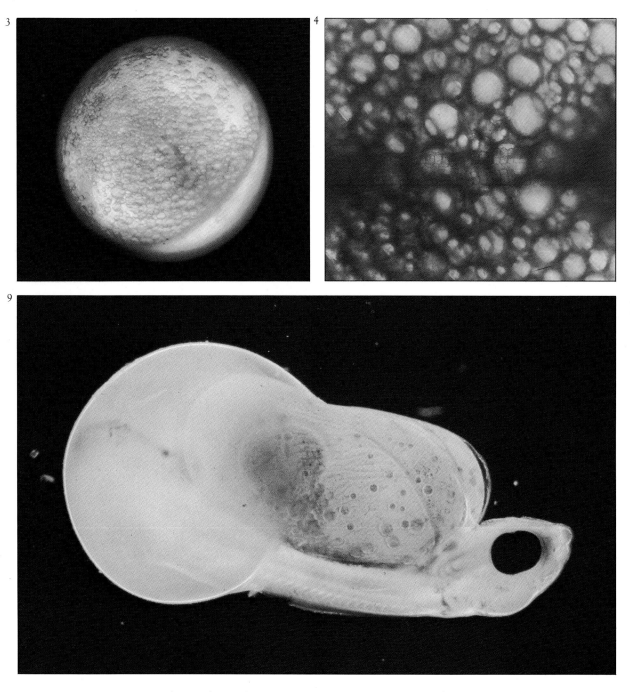

3

4

9

The development of a rainbow trout. (1) The egg, a few hours after fertilization, showing fat droplets. (2) Early cell division produces a disc sitting on the yolk. (3) The embryo can be seen as a faint streak running diagonally across the egg. (4) A closer view shows the two parallel rows of somites, the blocks of tissue which will become the trout's swimming muscles. (5) The brain is forming as a swelling at the front of the neural tube. (6) The embryo takes shape with a large head and curled body. (7) The eyes are becoming pigmented cups. (8) The lens is visible in the right eye. (9) The trout larva finally hatches out with its yolk sac.

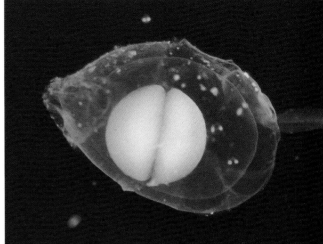

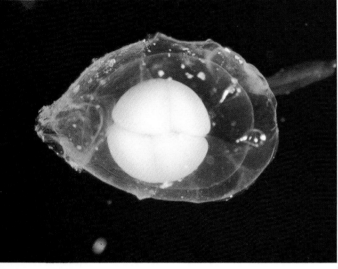

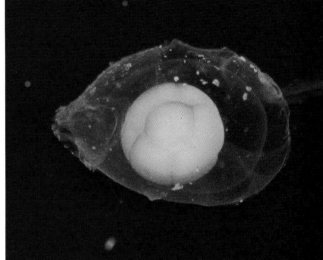

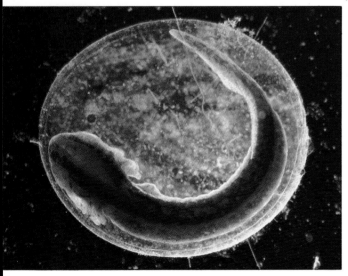

Development of the smooth newt, seen from above.
Top left: a new-laid egg surrounded by jelly.
Top right: the egg, after the first cell division.
Centre left: after two cell divisions there are four cells.

Centre right: at the eight-cell stage, the embryo is represented by the four smaller cells that are lying on the four larger yolk cells.
Left: a half-formed embryo showing the external gills and limb buds.

When the zygote of a frog starts to cleave, it divides into two cells. This step, and many subsequent ones, can be seen clearly with a hand lens if the eggs are obtained immediately after fertilization. The first sign that the zygote is about to divide is the appearance of a vertical cleft running through the sperm aster and around the egg to divide it into two cells. One cell will become the left side of the animal and the other the right side. By tracing the progress of cells in the early embryo, following them as they divide and divide again, it is possible for scientists to see how each cell formed during cleavage gives rise to a specific part of the later embryo.

Minutes or hours after the first division of the frog egg – the precise time depending on the water temperature – another cleft appears in the surface at right angles to the first, although still in a vertical plane. The embryo now looks like a peeled orange but with only four segments. If the frog embryo is cut in half at this point, it will grow into two tadpoles, but one will die because it has no digestive system. A human embryo will survive the loss of one cell at this stage and become a perfect baby. This is very useful because it means that a few cells can be safely removed from an embryo during the early stages of its development and the chromosomes studied for signs of destructive genetic defects, such as Down's syndrome.

The next division is horizontal, in the equatorial plane, and it divides the animal half of the embryo from the yolk. Cell division continues at a great rate: after the eight-cell stage, the embryo divides into 16 cells, then 32 and so on until it has divided 12 to 15 times and has become a mass of small cells forming the hollow blastula.

During the later stages of cleavage the cells may cease dividing simultaneously because the speed of division is affected by the amount of yolk in the cells. If there is not much yolk, as in starfish or sea urchin eggs, the cells continue to divide regularly. But if there is a mass of yolk, as in a frog's egg, cleavage is more rapid in the animal hemisphere and the cells become smaller than those in the yolky hemisphere which remain large. A similar pattern is found in molluscs. In insects and their relatives, cell division takes place only on the surface of the embryo to produce a layer of cells surrounding a yolky mass, and in most vertebrates, including sharks, reptiles, birds and the egg-laying monotreme mammals, a disc of cells forms on the animal hemisphere, leaving the rest of the egg as a yolky mass with no internal divisions. The eggs of placental mammals, including humans, have very little yolk and they form a ball of cells.

The cells produced by cleavage form the material – the bricks, so to speak – from which the basic shape of the future animal will be constructed in the next stage of development. This is the process of forming a gastrula, called gastrulation, and consists of actual movements of cells in the embryo so that they become arranged into tissue types. All animals above the level of sea anemones and jellyfishes have three types of tissues which are derived from three layers of cells formed by the radical reorganization that takes place in gastrulation. The outer layer, the ectoderm, forms the skin, nervous system and sense organs. The middle layer, or mesoderm, produces the muscles, skeleton, blood system and sex organs, and the inner layer, or endoderm, gives rise to the digestive system and lungs.

These three layers are formed as the blastula turns into a gastrula. In the frog, a curved slit called the blastopore forms just below the equator of the egg and the outer layer of cells on the animal hemisphere actively moves into the slit and back under the surface – imagine a conveyor belt rolling over its end pulley. Eventually all the pale yolk cells are enclosed except for a white spot, the 'yolk plug', contrasting with the now otherwise completely black embryo. The cells on the surface are destined to become the ectoderm; those that have moved inside the embryo form the future mesoderm and the yolk cells form the future

Right: drawings of a frog embryo sectioned vertically to show the transformation of a blastula into a gastrula. The outer layer of cells moves into the embryo through a slit called the blastopore and the yolk becomes totally enclosed.

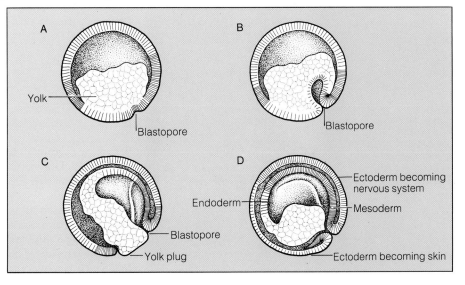

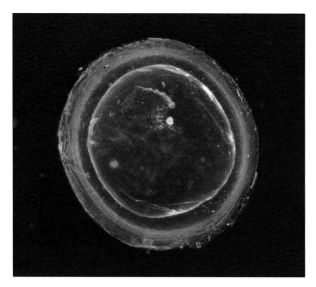

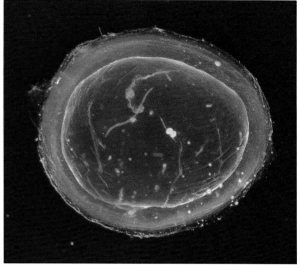

endoderm. The gastrula now rolls over so that the yolk plug, which continues to get smaller and will eventually disappear, comes to lie near the 'equator'.

The basic plan of the embryo, which was laid down when it was a single cell, is refined by gastrulation and regions are 'earmarked' to become parts of the adult body. As we have seen, part of the outer layer of cells or ectoderm on the upper half of the gastrula will develop into the nervous system and the remainder of the ectoderm becomes the skin. It may seem odd that the brain and nerves come from the outside of the embryo, but this is explained by the way that the nervous system arises after gastrulation. A thickened strip of ectoderm, the neural plate, rises in two parallel folds and bends inwards to form the neural tube. This is destined to become the spinal cord and three swellings at the head end mark the three main divisions of the brain. If something goes wrong and

the tube does not form properly, so that it remains open for part of its length, the result in humans will be the malformation called spina bifida.

Creating an organ depends on two processes. The first is induction in which the position of each cell determines how it will develop because it receives a series of chemical messages from the cells around it. When a cell divides, each daughter cell receives identical genes. However, the new cells receive a signal that activates only certain genes. For example, all human cells contain genes for the blood pigment haemoglobin but they are activated only in cells that will develop into red blood corpuscles. What many of these activating substances are and how they work is not yet fully understood, but it is known that they diffuse from neighbouring cells and from the embryo's environment. For instance, the alimentary canal and the various digestive organs develop from

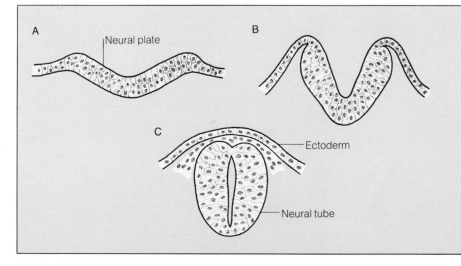

The development of the great grey slug. *Top far left*: in the middle of the transparent albumen that fills the shell, the zygote can be seen as a white spot. *Top left*: the first division produces two cells. *Top right*: the egg has little yolk and the embryo is nourished by albumen. A gland, resembling a golf ball, digests the albumen. *Top far right*: the embryo looks rather like a free-swimming

Left: the nervous system develops from a thickened ridge of tissue, on the top of the embryo, called the neural plate. The edges rise, fold over and fuse to form the neural tube.

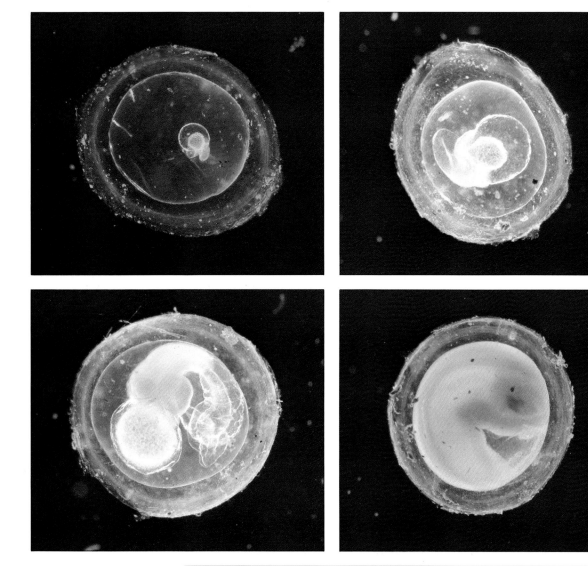

larva and has hair-like cilia which spin it inside the fluid-filled egg. *Centre left*: the digestive gland has grown and the embryo continues to rotate, although now the movement is caused by the beating of the tail. *Centre right*: the digestive gland and tail have disappeared and the embryo takes on the appearance of a slug. *Right*: the baby slug emerges from the eggshell.

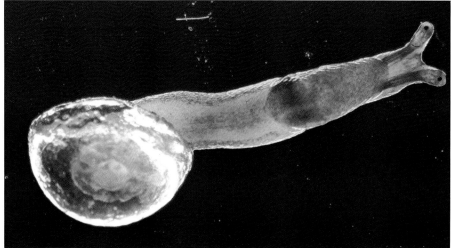

the tube of endoderm formed during gastrulation. The surrounding mesodermal tissue sends out signals that induce the intestine to form a side branch in the thorax region and this will become the lungs; another branch in the stomach region will become the liver.

The movements of cells in gastrulation and subsequent events are controlled by changes in the 'stickiness' of the contact between a cell and its neighbours. When moving, the cells slide freely over each other but eventually stick to form a solid mass of tissue when they have reached the right place. Contact with a particular type of cell increases the 'stickiness' and the 'join' becomes permanent.

Other major structures are forming at the same time as the alimentary canal. Some of the mesoderm forms the stiffening notochord and other mesodermal tissue forms blocks on each side of the notochord. These blocks or somites will become muscles and rudiments of the kidneys and blood system.

Meanwhile, in the endoderm, the simple tube which is the beginnings of the digestive system links with openings at the front and rear to make the mouth and anus, and on the exterior surface of the frog embryo rudiments of the gills and sense organs are forming. The embryo is now about a fortnight old; it is still only a half-formed animal – a tadpole – but it struggles free from the egg membranes and lies passively on top of the remains of the jelly. When its mouth opens, it starts to feed on minute algae and further growth and development proceed apace. Until this time, the energy for its growth and organization has come from the yolk, but this has become insufficient and is now supplemented from outside sources.

The embryos of animals as diverse as marine worms, insects, fishes and amphibians hatch out of the egg at an early stage to become free-living larvae that feed themselves. Others are given sufficient yolk for development to be completed within the egg and they hatch out as replicas of the adult. By doing so, they avoid the perils which beset a larva, and in some animals including the birds, reptiles and insects, the embryos have cut free from the ties of water by living inside their own aquatic environment. To make this possible the embryo requires a life-support system to provide it with food and oxygen, remove excess carbon dioxide and other wastes and regulate its water balance.

The early embryo of a bird is a disc of cells, two layers thick, sitting on top of the yolk. When a new-laid hen's egg is broken open the embryo can be seen as a small white spot 3–4 millimetres across. If the egg has been fertilized, the spot has a raised, translucent centre caused by the two layers of cells, the ectoderm above and endoderm below, being separated by a fluid-filled space. A few hours later, a groove known as the primitive streak runs down the centre of the disc, marking where cells are migrating into the space to form the mesoderm. The neural canal is the first organ to form. It appears as a pavement of cells that, as in the tadpole, throws up two ridges which fold over and fuse into a tube.

Until now the disc of cells has been able to absorb sufficient nutrition, through its lower surface, from the huge mass of yolk on which it sits, while oxygen and waste products have simply diffused in and out. This form of supply cannot keep up with demand as the embryo grows and it creates four membranes –

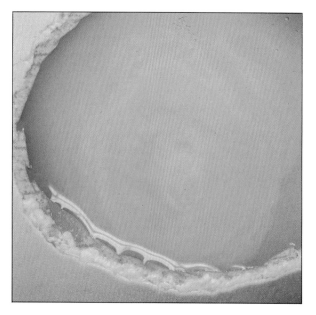

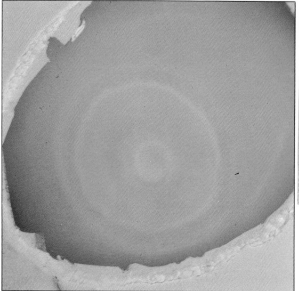

the yolk sac, amnion, chorion and allantois – to act as a life-support system.

After about two or three days, when the disc is 25 millimetres across, access to the yolk is improved by the edge of the ectoderm spreading outwards over the surface of the yolk and splitting into two layers with a space in between. The lower layer remains pressed against the yolk and becomes the yolk sac. At the same time, red spots, visible under the microscope, appear where blood corpuscles are being produced. A day later tiny capillaries form; they join up to make a network of blood vessels which connects with the heart. Blood is pumped round to pick up nutrients from the yolk and carry them back to the embryo.

The heart starts as a simple pulsating tube that causes the blood to ebb and flow, but in the space of a few hours it changes into the almost complete organ. Three bulges in the wall are linked by thickened sections that become valves to make the blood flow in one direction. The tube becomes S-shaped and the bulges fuse to become the chambers of the heart. As the lungs will not function for some time, the blood bypasses them and flows from the right auricle straight into the body's main artery via an opening called the ductus arteriosus. When breathing starts, the ductus arteriosus closes and blood flows through the lungs to pick up oxygen. Mammals have a similar system. The ductus arteriosus sometimes fails to shut; in humans this results in a 'blue baby' whose colour indicates that blood is not flowing through the lungs properly and is therefore not fully oxygenated.

Toxic nitrogenous wastes are formed during the manufacture of the proteins used to build the tissues, and they must be removed from the body. Aquatic

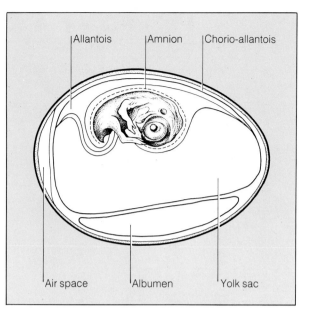

Above: diagram of a hen's egg, at about ten days, showing the membranes that form its life-support system. As the embryo grows, the airspace becomes larger and the albumen passes into the amnion so that the embryo can swallow it.

Opposite far left and left: a hen's embryo after one or two days of development is a spreading disc of cells sitting on the yolk. This can be seen when a fertilized egg is broken open. *Below left*: three days after fertilization the embryo has developed a network of blood vessels and a simple heart is pumping blood to collect much-needed nourishment from the yolk.
Below: nourished by the expanding blood system, the four-day-old embryo is taking shape and the internal organs are forming.

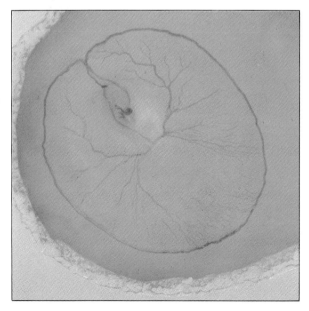

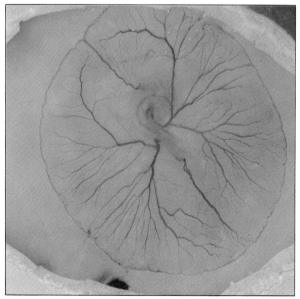

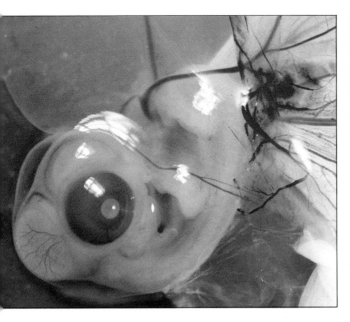

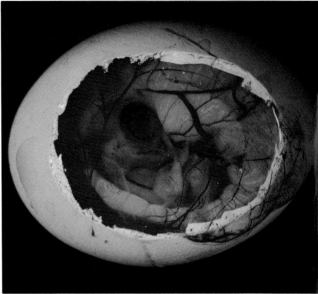

eggs get rid of them in the form of soluble ammonia which easily diffuses into the surrounding water. This is not possible for the terrestrial birds and reptiles, so they excrete uric acid which is filtered out of the bloodstream by the kidneys. These start to form only two days into development and they drain into the allantois, the membraneous bag which forms outside the embryo. Loss of precious fluid from the embryo is cut down by the water which flushes the uric acid through the kidneys being reabsorbed into the embryo's body, leaving the uric acid in the allantois as solid crystals. As the yolk and albumen are used up, the allantois expands into the space they leave and, when the chick hatches out, it is jettisoned with its store of toxic wastes.

To cater for the demands of respiration, which necessitates an inflow of oxygen and dissipation of carbon dioxide, the wall of the allantois fuses with the chorion membrane. This double membrane, called the chorio-allantois, rapidly grows around the inner surface of the eggshell and develops a system of blood vessels which acts as a lung for the diffusion of gases.

The young embryo's final requirement is a cushion to protect it from being jarred. When the embryo disc splits, the upper section rises on each side of the embryo and meets over it to form a fluid-filled bag, the amnion, that acts as a shock absorber and creates a watery environment.

With all systems functioning, the embryonic bird is set to complete its development and take on its final appearance by continuing differentiation and moulding of tissues. By the sixth day of incubation, most of the organs in a hen's egg are taking shape. The head with its enormous eyes is the most obvious

feature; the body has folded to bring the beating heart to its final position and the gut is forming. The limbs are also showing as tiny buds. At ten days, the embryo is recognizable as a bird with beak, wings, feet and even tiny spots on the skin which represent the feathers. The fluid-filled amnion now takes up about half the space inside the egg, at the blunt end, and the albumen gathers at the narrow end. The embryo is lying on its back across the egg's axis, but very shortly the tail will press into the narrow end and the head will be tucked between its legs.

The complicated processes of organ formation, as tissues change position and take on their specialized functions, are not yet fully understood, but a description of how the eye and the limbs are formed will give an idea of this vital part of an embryo's development.

The eye has two main parts: the light-gathering apparatus of cornea, iris and lens throws an image onto the sensory system of the retina. Yet this intricately structured organ of muscles, nerves, glands, blood vessels and transparent tissues, from which the brain is able to construct a wonderfully detailed interpretation of the animal's environment, is formed from a simple sandwich of basic tissues.

The embryological origin of the eye lies in the formation of the neural tube. The parts that will become the brain are induced by substances from neighbouring tissues to send out two bulb-shaped optic vesicles. These swell until they make contact with the layer of ectoderm forming the 'skin' of the embryo. Each vesicle has a single layer of cells and it collapses inwards, like a tennis ball being pushed in, to make a two-layered optic cup. The inner layer will become the retina with its light-sensitive rods and cones. The outer layer will fill with pigment and the front of the optic cup becomes the iris diaphragm that controls the amount of light entering the eye.

The development of the eye has become a classic textbook example because of the way differentiation is controlled by induction. The shaping of the optic cup is controlled by a two-way process of induction involving both the ectoderm sheet overlying it and the surrounding mesenchyme (the part of the mesoderm that is destined to become connective tissue). At the same time the optic cup induces the overlying ectoderm to form the lens. This starts as a thickening of tissue over the optic cup, then folds inwards and 'nips off' to make a hollow ball called the lens vesicle, which eventually loses its cellular structure to become transparent. (Studies of lens formation have shown how blindness in infants can be caused by the mother catching German measles early in pregnancy. The virus responsible for the disease can only attack the lens while it is unprotected. After it has sealed off as the lens vesicle, at about the sixth week of pregnancy, it is safe.)

It is now the turn of the lens to act as an inducer. It controls the formation of the cornea, the transparent 'window' of the eye, out of the ectoderm sheet from which the lens itself was derived. The cornea becomes transparent because pigment and gland cells, which are found in other skin tissues, are prevented from forming.

Experiments show that the construction of the eye is controlled by chemical messages that pass between optic vesicle, lens and corneal tissues, and the interactions between them are typical of the way organs form. The development of the limbs proceeds by essentially the same mechanism of two tissues interacting with each other.

A limb bud arises from mesoderm cells accumulating under the ectoderm and inducing it to thicken into a ridge. At this stage the fate of the cells

Above left: a seven-day-old hen embryo inside the tight-fitting amnion. The eye has developed a lens and the limbs are forming. Note the blood vessels running to the yolk. *Above right*: the membranes have been parted to show a 14-day-old embryo. From its original position, lying on its back, the embryo has rolled over to lie on its left side with its tail pointing towards the narrow end of the shell.

Right: a brood of red jungle fowl immediately after hatching. The most recently hatched chick still has the allantois, containing the excretory products, attached to its body.

109

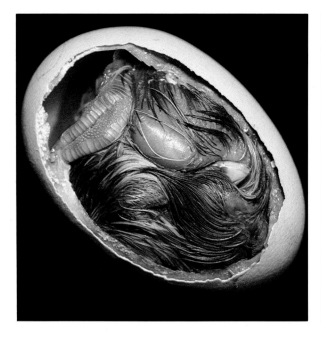

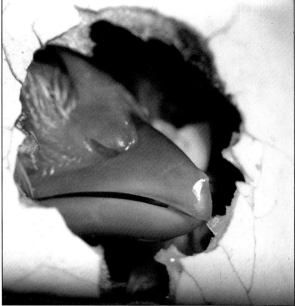

has already been decided: specific clumps of the mesoderm cells are destined to become particular bones. The limb now begins to grow out from the body with new tissues forming at the tip like a branch growing from the trunk of a tree. In a leg, the femur forms first, followed by tibia and fibula and finally the bones of the ankle and foot. Muscles and tendons are produced at the same time. When all the parts of the limb are in place it assumes its final shape by a kind of embryological sculpting, in which certain cells die. For example, the foot is transformed from a simple paddle into separate toes by the tissue between them dying.

Other organs in the embryonic chick similarly arise from simple structures that take on the adult appearance, but some are never completed. The rudiments of gills appear and vanish; a tail with vertebrae and muscles forms and is withdrawn to the remnant stump that will support the tail feathers of the adult bird. And so the organization and reorganization continue, from egg to animal, following a course that is predetermined by heredity and is approximately the same for all species except in final details. But it is still not known how one egg becomes a frog and another a hen, nor for that matter how a horse comes to have a leg at each corner.

The concluding stages of development in the egg are concerned with growth and adding the finishing touches. A chick taken from its egg before it is due to hatch may still survive, as a premature baby will survive with the proper care. But even a full-term hatchling chick is incomplete: its reproductive organs are undeveloped, its plumage is rudimentary and

some of its behaviour remains latent. A young bird takes to the air at a specific age, not because it has been practising nor because its muscles have developed, but because the nervous system has grown and made the appropriate links with the muscles.

Some bird species hatch out at earlier stages of development than others. A hen's chick can run and feed itself soon after hatching, but a newly hatched sparrow or robin is naked, its eyes are closed and it is helpless. Almost its only reaction is to lift its head and open its mouth in response to the vibrations that indicate the arrival of a parent with food. It is really little more than an embryo which, like the frog's tadpole, has emerged from the egg because it needs food to supplement the remains of the yolk. The period from fertilization to hatching is not an isolated chapter in the life of an animal but a stage in the journey from zygote to fully grown, mature adult. The egg is a protection until the animal is sufficiently well developed to cope with its environment. This comes early in development for a sea urchin but late for a bird.

Extra embryos

Identical twins are the result of a single fertilized egg developing into two embryos. In humans and most other animals, such twins, and larger numbers of identical offspring, are rare 'mistakes' in the early embryological process. Fraternal or non-identical twins are the result of two eggs being liberated into the oviducts at the same time and 'double-yolked' eggs have the same origin. A few species have

Far left: the perfect package —
a few days before hatching the
chick occupies almost the
whole of the egg. It has twisted
its head under its right wing so
that its beak faces the blunt end
of the egg.

Left: incubation is complete.
After 21 days the chick is ready
to emerge into the world.

Right: a gosling photographed
through its eggshell. The wing
can be seen folded over the
beak and the yolk sac is lying
behind the head and foot.

Below: hatched out but not
fully developed, this magpie
nestling is little more than an
embryo. Its eyes are closed and
its body is naked.

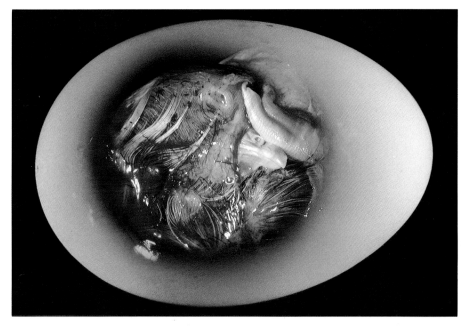

established the production of many identical embryos, called polyembryony, as the normal method of breeding. For example, the nine-banded armadillo of Mexico and the southern United States gives birth to litters of identical quadruplets derived from a single fertilized egg. Why it should behave in this manner is not known but presumably it is a quirk of evolution.

Many of the parasitic wasps – including the chalcids and ichneumons – that lay their eggs in the bodies of other insects, have developed polyembryony in a trend that reverses the usual behaviour of parasites. Because of the difficulties involved in finding their hosts, parasitic species tend to lay large numbers of eggs. The solution adopted by these insects is to lay one egg, or sometimes a few, in a single host insect and for each egg to develop into a large number of larvae. The record is held by the chalcid wasp *Litomastix trunctatellus*, whose single egg develops into around 1,000 identical larvae in the caterpillar of the silver Y moth.

The formation of multiple embryos takes place during cleavage, when individual cells or groups of cells become separated. To provide food for this growing brood, the polar bodies which are normally shed and lost (see page 18) are retained on the outside of the egg to form a sheath of tissue that acts as a kind of placenta, absorbing nourishment from the host tissues and passing it to the embryos.

In Chapter 4 it was recounted how *Trichogramma* tries to avoid laying its egg in a host egg that is already parasitized because there may be insufficient food for two larvae. With one egg turning into many larvae, it is easy to imagine that competition within the host could become even more intense. This problem has been solved by the wasp *Copidosomopsis tanytmemus*, which lays its eggs in moth caterpillars. Each egg develops into 200 larvae which, although genetically identical, form two distinct 'castes'. One caste develops as 'soldier larvae' equipped with strong jaws. These wander through the body of the host which becomes a battlefield as they attack and kill any other parasite larvae they find. The soldier larvae have no excretory, respiratory or reproductive organs and, once their task has been fulfilled, they die, leaving the field clear for their slower-growing but anatomically complete siblings which develop unhindered and emerge as fully formed adults.

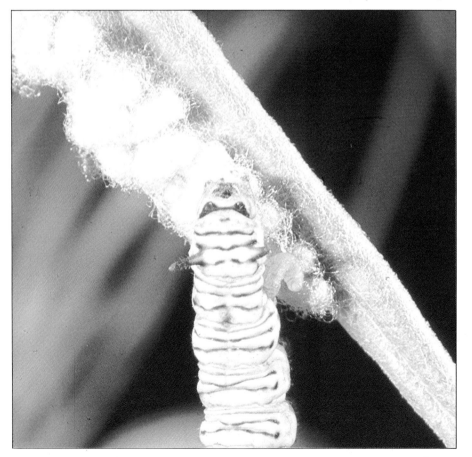

Left: an ichneumon larva emerging from the caterpillar of a monarch butterfly; other larvae have already made cocoons on the leaf. Many ichneumons lay an egg that develops into numerous larvae.

Right: these 'Siamese twins' developed from a single trout egg. A mistake occurred early in development that resulted in most organs being duplicated. If the mistake had happened earlier, the result would have been identical twins.

CHAPTER 8
Hatching into a wider world

Throughout its development, the egg has been protected by its enclosing shell or membrane, which has formed a barrier against a hostile environment. But when the time comes for the young animal to emerge, the situation changes: the shell becomes an obstructive barrier. To hatch out from its cradle, the young animal requires special mechanisms and if these fail, it can be trapped and doomed. When it does emerge, it faces the trickiest moment of its life: it leaves the sheltered environment of the egg and has to face the vagaries of heat and drought and the attentions of predators. Hatching is therefore a major event in the life of an animal, as important and traumatic as birth in animals that bear live young. To hatch successfully, the young animal requires a programme of instructions from its genes to co-ordinate the sequence of events, as well as signals both from within the egg and from the external environment to ensure the correct timing of hatching. Little is known about the nature of these signals, however.

The hatching process of birds has been well documented, since a successful outcome is a matter of economic importance for breeders of poultry, game and ornamental birds. Among other animal groups hatching is coming under scrutiny as commercial breeding, for instance of fish, and captive breeding for conservation of endangered species, become important.

The first outward and visible sign that a hen's egg is about to hatch is a tiny eruption – the 'pip' – on the surface of the shell near the blunt end; but preparation for hatching has begun some time earlier. The embryo has been making small movements since the early stages of development, but these have been uncoordinated jerks and twitches as muscles contracted spontaneously. During the last few days of incubation these contractions are replaced by co-ordinated movements which prepare the embryo for hatching. Some five days before 'pipping', a domestic chicken starts the hatching process by swallowing the remaining fluids that surround it, and

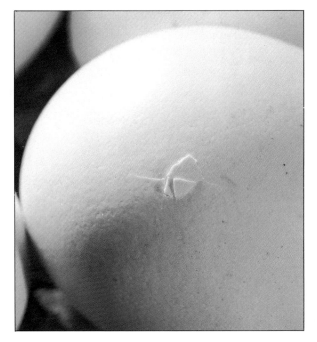

Opposite left: a jungle fowl egg has just pipped. The chick inside has cracked the shell with the egg tooth on the tip of its beak. When the hole is fully open, the chick will begin to breathe properly.

Opposite right: a moorhen chick has made a ring of holes to cut the top off the shell and it will soon push its way out.

Right: a silver pheasant chick, having cut a ring of holes around most of the circumference of its shell, has pushed off the top like a hinged lid. It is now pushing with its feet to struggle free of the shell.

Below: a mallard duckling has finally worked its way out of the eggshell and is taking an interest in its surroundings.

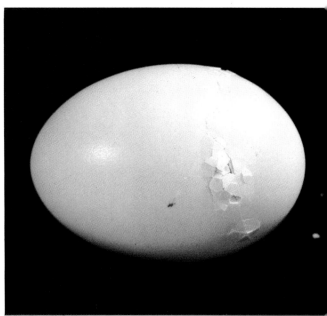

A pigeon egg hatches. Seventeen days after laying, the egg pips as the chick's beak makes the first crack in the shell. The chick revolves inside the egg and continues to chip the shell to produce a ring of cracks that almost separates the top of the shell.

Now it can heave against the two sections of the shell to push them apart. First a wing appears, then a foot. The chick uncurls and head and neck emerge. Having struggled almost free of the shell, the chick rests while the remains of the egg fluids dry.

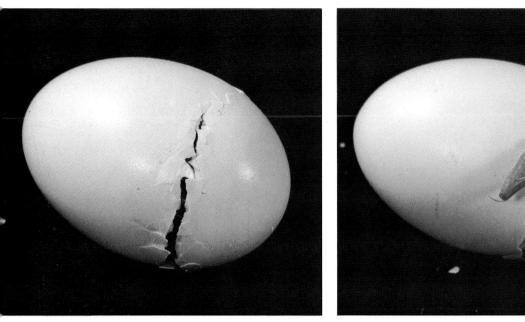

gradually shifting its position within the egg. This is called tucking. It brings its head from between its legs, twists its body and tucks its head under its right wing, so that it faces the blunt end of the egg. Occasionally it even flaps its wings – as well as it can within the confines of the egg.

As the embryo grows to its full size its metabolism produces too much carbon dioxide to diffuse through the allantois (the membrane that acts as the embryo's lungs) and the pores in the eggshell, so the carbon dioxide level in the tissues rises. One result is that the yolk is drawn up into the embryo's body, where it continues to be digested until after hatching, when the navel closes over it and heals. The build-up of carbon dioxide also causes a twitch in the powerful 'hatching muscle' at the back of the chick's head. This levers the head backwards and forces its beak into the airspace at the blunt end of the egg, which has gradually expanded as water has been lost by evaporation. Breathing with the lungs can now begin, but the airspace is not very large and direct contact with the atmosphere is needed. The airspace becomes foul as its carbon dioxide content rises; this stimulates even stronger twitches in the hatching muscle and the egg tooth or caruncle (a horny lump on the beak that disappears after hatching) is pounded against the shell. During the later stages of incubation the shell has been weakened as calcium and other minerals were removed to help build the embryo's skeleton, but eggshell is like bricks and mortar – it is very strong in compression but has no tensile strength. Because of its shape, an eggshell resists compression; it is very difficult to squash an egg by squeezing and an ostrich egg can withstand the weight of two adult humans standing on it! But only slight pressure from within sets up enough tension in the shell to crack it. Repeated twitches of the hatching muscle make small cracks in the shell which push up the characteristic pip. Eventually the broken slivers of shell fall away, leaving a hole through which can be seen the chick's beak and its white egg tooth.

Once the egg has pipped, the chick has a plentiful supply of fresh air. The allantois is no longer required for exchanging gases and it dries up. At the same time blood begins to circulate through the lungs and they expand, filling with air and commencing their proper function. After the intense effort of pipping, the chick rests for a few hours, or even days, before the final stage of hatching which takes less than an hour.

When it is ready, the chick cuts itself out of the egg. Continuing from the first pip, it uses its egg tooth to make a neat row of connecting holes around the circumference of the large end of the egg. After each crack has been made the chick relaxes, then, pushing with its right foot, it twists a few degrees

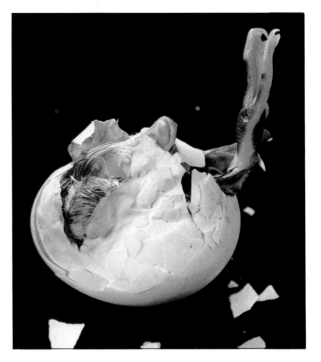

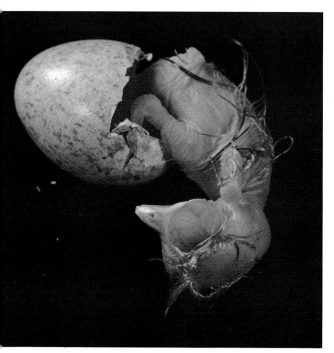

anticlockwise so that its bill is positioned to make the next hole. The success of this crucial task depends on the chick having adopted the proper posture. If it has not got its feet into the pointed end or its bill into the correct position, it will be unable to hatch and will die imprisoned in its shell. Exceptionally, as in the case of the Muscovy duckling illustrated on page 118, the chick is strong enough to force its way out of the shell even when it is wrongly positioned. Contrary to what is often believed, birds rarely, if ever, help their chicks out of the egg. Some parent birds pick pieces from the eggshell, but this is probably not enough to help a trapped chick. Ostriches, however, have been known to pull their chicks out of the shell.

The amount of pipping necessary to weaken the shell sufficiently for the chick to escape depends on the strength of the shell. Chicks of hens and ducks rotate about two-thirds of a complete circle, while ostriches with more brittle shells cut less than a quarter of a circle, and owls and pigeons with tougher eggshells make a complete ring. Eventually the chick forces the cap off by pushing with its feet, heaving with its shoulders and straightening its neck. Once its

Above left: eggs usually fail to hatch when the chick is lying in the wrong position. This Muscovy duckling is lucky because it has managed to force its way out by kicking with its strong clawed feet.

Below left: a greater flamingo appears to be helping its chick out of the egg. It is more likely that it is eating fragments of shell for its calcium content.

Above: a blackbird kicking itself free of the shell in the final stage of hatching.

Right: newly hatched lapwing chicks close their eyes to improve their camouflage. The white egg tooth can be seen clearly on the tip of the bill.

head is free, it quickly escapes from the egg, using its wings and kicking the shell away with its feet. After this supreme effort it may lie exhausted on its side for a while, still curled, but it later lies belly-down with its head and neck stretched out.

While the chick is lying in this position its down dries. In time its neck strengthens so that it can raise its head and its legs gain sufficient strength to support its weight. A domestic chick may stay in the nest for several days after it has forced its way out of the egg, sustained by the remains of the yolk which still makes up a quarter of its weight at hatching. The yolk is especially important for chicks that feed themselves; songbirds, which stay in the nest for much longer and are fed by their parents, hatch with only a small supply of yolk.

There are few variations among birds in the basic pattern of hatching. Ostrich chicks lack an egg tooth but they are strong enough to break out without it. Chicks of the woodcock and willet (both members of the sandpiper family) have an egg tooth on each half of the beak; they do not cut a ring of holes before hatching but push the bill through the pip and heave, splitting the shell in a mass of untidy cracks which

make the hatched egg look as if it has been trodden on. However, a pigeon chick also has two egg teeth, but it opens the egg in the neat, sawn-around manner of the one-egg-tooth chick and the function of the lower egg tooth is not known.

The chicks of the Australian megapode family (see page 142) hatch out without the assistance of beak or hatching muscle. These birds have reverted to a reptile-like habit of laying their eggs in pits and their method of hatching also resembles that of reptiles. The embryos start to develop an egg tooth but it disappears before it is complete and the hatching muscle never develops. Megapodes hatch out at a very advanced state of development and are independent of their parents from the moment that they emerge. This means that they are strong enough to force a way out of the unusually thin eggshell without first making a ring of holes. The embryonic megapode remains with its head between its legs in the sharp end of the egg. By bracing its back against the shell and stretching its legs and wings, it simply bursts the shell open and struggles out.

If eggs are incubated in isolation, rather than in a clutch, they will hatch within a day or so of the usual

incubation period of the species. This suggests that there is a programmed mechanism in the embryo which regulates its hatching. In some species this 'clock' is overridden by signals from outside the egg. A clutch of a dozen mallard eggs will hatch within the space of around three hours. Co-ordination at this stage is clearly important: a duckling that hatches out after its siblings runs the risk of being left behind when their mother leads them from the nest, never to return. Indeed, co-ordinated hatching is a feature of many species of precocial birds which leave the nest shortly after hatching. Clutches of bobwhite quail eggs, for example, hatch within the space of one or two hours. The eggs start to develop together because the adult does not begin to incubate them properly until the clutch is complete, but the speed of development can vary through unequal incubation of eggs in a large clutch. Bobwhite quails hatch out after 23 days' incubation and pipping starts about two days earlier. Some time after pipping, when breathing has started, a faint sound called 'clicking' can be heard from a bobwhite quail egg held close to the ear. If the eggs are touching, the embryos can hear each other's clicks and those from the first egg to pip stimulate the other embryos to speed their development until the whole clutch is clicking. Then they all go through the final hatching process at the same time.

Other bird species make a virtue of staggered hatching. Unlike the mallard ducklings and quail chicks that feed themselves, the young of altricial birds that stay in the nest as nestlings require their parents to invest considerable time and energy in bringing them food. When food is in short supply, the nestlings may become undernourished and their lives are placed in jeopardy. In such circumstances a staggered hatching of the clutch is an aid to breeding success. The owls, birds of prey, herons and swifts, among other birds, start incubating with the laying of the first egg. Each egg begins to develop as it is laid and the eggs hatch at intervals in the order in which they were laid. The result is that the older nestlings are larger and stronger than their younger siblings and consequently take the first places in the queue for food. Provided that food is plentiful the younger nestlings get fed eventually and the whole brood survives. However, these birds live on food that is subject to sudden fluctuations and the nestlings can go hungry. The older nestlings beg more vigorously

Left: a clutch of silver pheasant eggs hatches; some chicks are already out and dry, while others are about to start hatching. They will all be strong enough to leave the nest together.

Right: barn owl eggs hatch at two-day intervals in the sequence in which they were laid. The oldest nestling is nine days old and one egg has yet to hatch.

Above: a starling removes an empty eggshell from its nest. Getting rid of the shells is a safety measure when the young birds stay in the nest.

Below: partridge chicks leave the nest soon after hatching. There is neither the time nor the necessity for the parents to remove the shells.

than their younger siblings and push them aside, so they get the major share of such food as the parents can bring. The youngest weaken and eventually die. Thus the older nestlings survive at the expense of the younger ones, but the parents have a chance of rearing at least some healthy young, whereas they might lose the whole brood if food was shared equally.

When the young bird has struggled out of the egg, the remains of the once all-important shell are usually removed by the parents. Some birds eat the shells for the calcium, but the majority pick them up in their beaks and carry them from the nest, dropping them at some distance. This is partly to keep the nest clean and to prevent the nestlings being injured on the sharp edges of the shells, but ground-nesting birds such as plovers and gulls remove the shells because the white interior is very conspicuous and attracts predators. Others such as ducks and pheasants leave the empty shells in the nest when they walk away.

The hatching process has been studied in few animals other than birds, so only a sketchy summary is possible. It appears that none has such a complex pattern of hatching behaviour. This may be because the strength of a bird's eggshell requires specialized movements to break it open. Other animals can employ various swimming or respiratory movements which will be used in later life. Birds are unique in

Above left: a spur-thighed tortoise begins to hatch out beside the shell of another egg. Normally the eggs are buried and the hatchlings dig their way to the surface.

Above right: hatching is not a complicated process for a tortoise. It uses its egg tooth – a sharp spike on its snout – to penetrate the eggshell, then it simply bursts its way out.

Right: this grass snake has pierced its eggshell with an egg tooth protruding from its upper lip. A miniature version of its parents, it is looking out, ready to emerge if the coast is clear. If it is disturbed when almost out, it can withdraw into the shell.

that their hatching movements are never repeated after they have left the egg. Even reptiles, which are the direct ancestors of birds, lack specialized hatching movements. Crocodiles, tortoises and turtles have an egg tooth or caruncle which is a horny lump like that of birds, but lizards and snakes grow a real tooth which curves forward from the centre of the upper lip and is also lost after hatching. Pythons use their egg tooth to slash slits in the tough leathery shell, while lizards and crocodiles use theirs to pierce the shell with thrusting movements of the head. Once the shell has been breached, the young reptile struggles out with the same pattern of movements that it will use for locomotion. The two egg-laying monotreme mammals of Australia, the platypus and spiny anteater, follow the reptilian pattern of hatching. They have a true egg tooth and a horny caruncle; both are used to cut a hole in the shell, then the young animal struggles out with writhing motions of the head and forelimbs.

Fish and amphibians escape from the egg membranes partly by weakening them with enzymes secreted from special glands on the snout of the embryo and partly by vigorous swimming movements. One known exception is the rain frog which hatches as a froglet rather than as a tadpole. It has a tiny egg tooth on its snout which is used for rupturing the rather tough membranes.

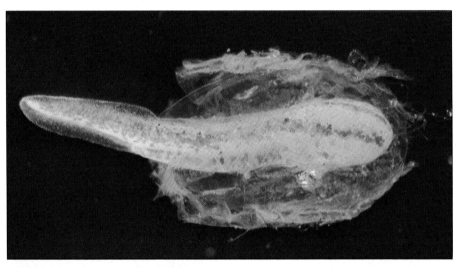

Left: the tadpole of a smooth newt has dissolved the egg membranes and is making swimming movements to get clear of the remains.

Right: rain frogs of tropical American forests hatch out as froglets rather than as tadpoles. Unlike other amphibians they have an egg tooth for cutting their way out of the egg.

Below: a trout leaves its egg; it is carrying the yolk sac that will nourish it for several days.

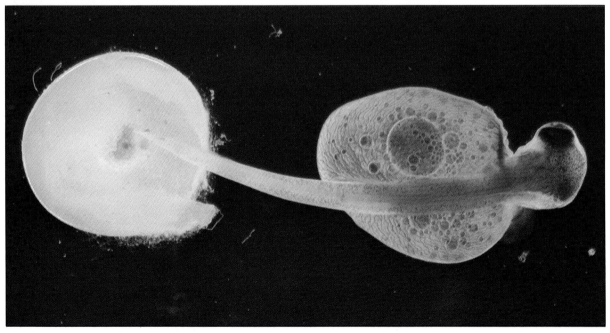

A number of invertebrates possess sharp spines for cutting the eggshell or membrane. Some insects crawl around inside the shell so that the spine acts as an internal tin opener, slicing a neat slit through which they wriggle. When the full-term mosquito larva starts to wriggle, its head is forced into the end of the shell where its hatching spine tears a slit. Other invertebrates hatch out by swelling up until they can no longer be contained and the membranes burst. This can be achieved by taking up fluid or air, but assassin bugs (whose name derives from their predatory habits) fill a special bag with gas. When it is full it forces off the lid of the egg, then the bag explodes to leave the way clear for the bug to emerge.

Finally, the invertebrates provide one of the few known cases of a parent helping with the hatching. The American lobster carries her eggs glued to her swimmerets. When hatching is due, she raises herself up on the tips of her legs and beats her swimmerets so violently that the egg membranes burst and the embryos are shaken out. It is not known what triggers this behaviour – that is, how the lobster 'knows' when her young are ready to emerge – but eggs taken from the parent rarely hatch.

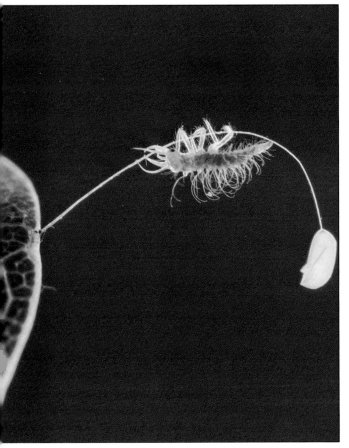

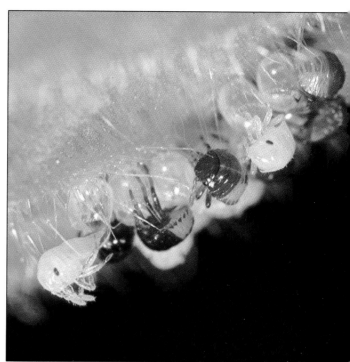

Top left: the caterpillar of a red admiral butterfly newly emerged from the egg. As part of the hatching process it has swallowed air and blown itself up; this is why it is larger than the egg it came out of.

Top centre: swallowtail moth caterpillars chew neat holes in their eggshells so they can crawl out.

Top right: eggs of the woundwort shieldbug have a line of weakness around the top so that a lid pops off when the young hatch out.

Far left: a lacewing egg on the tip of a long stalk is safe from predators, but the larva has to climb down the stalk to the leaf where it will feed on aphids.

Left: newly hatched praying mantises are tethered by silk threads as they emerge from the foam capsule that protected the eggs.

Right: Roman snails hatch by dissolving part of the eggshell. Their first task is to eat the remainder of the shell, which provides more calcium for their own shells.

CHAPTER 9

Parental care

Most animal species lay their eggs and abandon them to their fate. Parental investment in their survival is limited to little more than laying them in a situation where they may be relatively safe from predators and the vagaries of the environment, and where the emerging young will have a reasonable supply of food. Some animals improve on this casual behaviour by making a nest to hold the eggs and a few provision the nest with food for the emerging young. Only a minority of animals have taken further steps in parental care and stay with their eggs to tend them and improve their chances of hatching.

The advantages of parental care are clear. An animal's biological success is measured by the number of viable offspring it produces. Laying a large number of eggs most of which are destroyed by accident, disease or predation is not being successful. But caring for the eggs and preventing these losses will improve the parents' biological success.

The question this immediately raises is, why don't all species care for their eggs? The simple answer lies in the need to balance costs. Parental care costs time and energy; it may even cost the animal its life if, for instance, it defends its eggs so stoutly that it, too, falls victim to a predator. It may be cheaper, in terms of the use of precious energy resources, simply to produce large numbers of eggs and to risk losing most of them. This is an extension of the argument developed when discussing the relative advantages of laying large and small eggs (see Chapter 6). Each case has to be viewed on its own merits; some species have opted for large eggs and perhaps parental care, others for mass production of small eggs. The benefits and costs that accrue either way are often very subtle and not obvious without detailed investigation of the animal's ecology. For example, it might be found that carrying the eggs hampers the parent to such an extent that it cannot feed properly, and it has been shown that one starfish runs the risk of being swept off the rocks when carrying eggs.

Although a relatively small number of animals care for their eggs, they are found throughout the animal kingdom and are by no means confined to the 'higher' animals such as the birds. A surprising number of

invertebrates establish some form of relationship with their eggs and some species show quite complex maternal behaviour.

The familiar earwig has a simple form of maternal behaviour. The sexes pair up in autumn and pass the winter together in a chamber excavated in the earth. In spring the male, recognizable by his curved pincers, departs and the female remains to lay her eggs. She gathers them into a pile and gives them her undivided attention until they hatch, then she stays with the hatchlings until they are fully grown. During this period the brooding earwig spends a considerable time picking up her eggs and licking them to remove fungi and bacteria. This is thought to be the function of parental care in other insect species, as well as in fish and salamanders, where cleaning the eggs appears to be the only expression of parental behaviour.

It may be that even a poorly armed animal such as an earwig can protect its eggs by aggressive behaviour. Parasites – for example, the chalcid wasps and ichneumon flies – are probably the most serious threat. They would find it difficult to carry out the necessary assessment of the egg and the delicate process of inserting their own egg while being harassed by a parent. It is perhaps to this end that some bugs lash out at egg predators with their legs. Many species of ground beetle dig a pit in which they lay their eggs, but in one group, the tribe Pterostichini, the females of some species remain with the eggs until they hatch and the larvae disperse. It seems that this is an adaptation to the cooler climate of the mountainous regions where the eggs take longer to develop and thus would be at greater risk from egg predators without parental care.

An alternative to depositing the eggs in a nest is to carry them, as has already been described for marine and freshwater crustaceans such as lobsters, crabs and crayfish, which carry eggs glued to the outside of the

---

One of the simplest forms of parental care is to choose a safe place for the eggs. This fish louse, shown with its pair of egg sacs, lives on the skin of sea fishes such as bass but has to leave its 'home' to lay its eggs on stones.

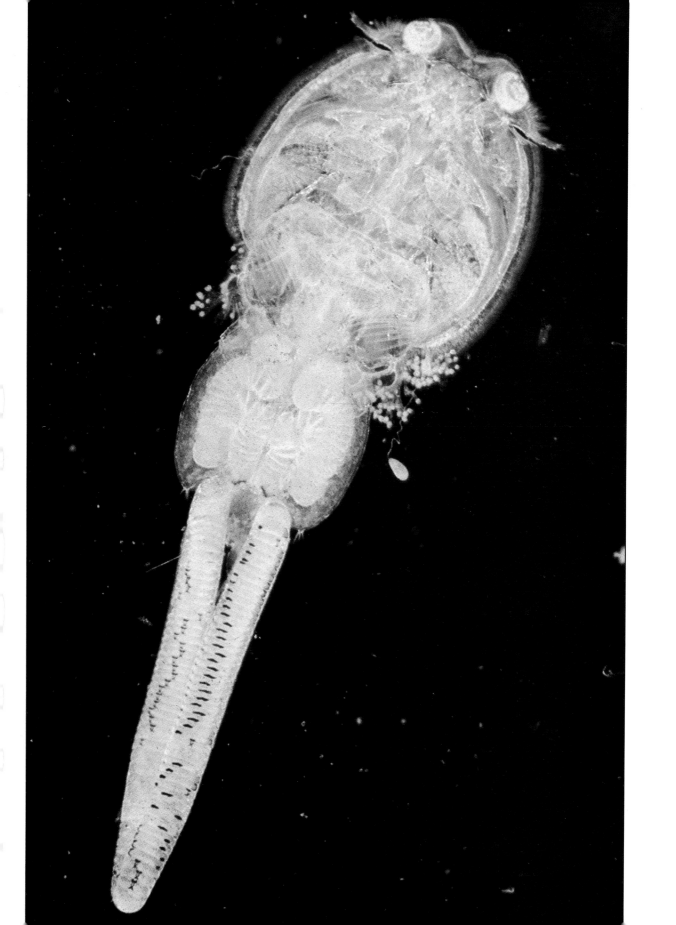

 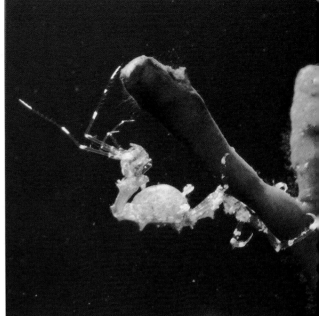

body, and water fleas that carry the eggs inside the carapace or in special brood pouches. Similarly, the rough winkle of European shores has developed the ability to carry its eggs: the gland that in other species secretes the protective jelly coating for eggs has been modified into a brood chamber where the eggs are kept until they hatch. This enables the rough winkle to breed at the top of the shore and in saltmarshes where exposed egg masses would perish from desiccation or could be smothered in silt. The drawback is that fewer eggs can be laid because of limited space in the chamber, but this is presumably outweighed by the advantage of the increased survival rate of the eggs.

Octopus eggs are attached in strings to the rocks that form the female's sea-bed lair. They benefit from the presence of a fierce, predatory mother, but she also cleans them by caressing them with her arms and blowing jets of water over them. Wolf spiders are also fierce, protective mothers. These spiders actively chase prey, rather than snaring it in a web, and the females clutch their egg cocoons to their bodies. If the cocoon is prised away, the spider rushes about searching for it but will be content if a small pebble is substituted in its place. From observations of cocoons removed from the mother and left on the ground, it seems that the main function of carrying the cocoon is to protect the eggs from moulds.

Another advantage of carrying eggs is that a safe hiding place is likely to be too cold or damp. Carrying the eggs not only gives them protection but allows the parent to shift the eggs to favourable places. The

pirate spider *Pirata piraticus* is a wolf spider that lives deep among sphagnum bog moss in wet places. It will die if taken away from humid, cool conditions, but the females exhibit a streak of altruism by climbing to the tops of the bog moss stems to expose their egg cocoons to the warmth of the sun and so speed their development.

Among the fishes there are examples from across the whole range of parental care, from the cod which spawns in a school, shedding eggs and sperms into the water then paying no further attention to its offspring, to some sharks that give birth to live young, having nourished the developing embryo by means of a placenta. As with the laying of large eggs, parental care is most common among fish that lay their eggs in the uncertain conditions of the seashore or in fresh water.

Simple parental care is shown by the blennies that live in shallow seas and can be found among stones and seaweed in rock pools. Their eggs are glued to the insides of empty seashells or are hidden in rock crevices and the male (or, occasionally, both the male and the female) remains with them. Although this behaviour is often called 'guarding', there is no proof that the blenny ever offers active defence against an egg predator. The term is a relic of the language used by naturalists a century or so ago; it was used continually until the blennies' behaviour was examined properly, when it was found that there is little that these fishes can do to defend their eggs. The main purpose of the parental presence is to aerate the eggs by fanning them with their tails. The same is true

Far left: the mass of eggs on its swimmerets hinders a common prawn's movement, but this simple parental care greatly improves the eggs' chances of survival.

Left: the skeleton shrimp *Caprella acanthifera* carries her eggs in a dome-shaped brood pouch.

Right: the wolf spider *Pisaura mirabilis* carries her egg cocoon in her jaws. When the eggs are ready to hatch, she spins a dense nursery web and stands guard over the spiderlings.

Below: a meadow spider, *Lycosa amentata*, carries her newly hatched young. The empty egg cocoon is still attached to her silk-secreting spinnerets at the rear of her abdomen.

for the lumpsucker, another seashore fish. When the tide is in, the male fans the eggs with his pectoral fins, but at low tide he clings to the exposed rocks with a sucker formed from the pelvic fins. This is suicidal behaviour because he can be plucked off by gulls and crows or torn off and battered on the rocks by storms. One of the South American characins, *Copeina arnoldi*, has a novel way of avoiding aquatic egg predators: it lays its eggs out of water. Both male and female leap out and cling, belly uppermost, to the underside of a leaf for a few seconds. Eggs are rapidly laid and fertilized and the pair drop back into the water. The performance is repeated until all the eggs are laid, then the female departs. The male remains underneath the leaf and periodically splashes water over the eggs with his tail until they hatch one or two days later.

From simply 'guarding' the eggs, some fishes such as the salmon and stickleback have advanced their parental care by building nests. Others have taken to transporting their eggs. Among the banjo catfishes of South America, the male of one species rolls in its mass of eggs so they become attached to its skin, while during the breeding season the female of another species develops spongy tentacles on her belly to which her eggs become attached.

The most specialized of the egg carriers are found among the hundreds of species of fishes in the Cichlidae family. One group, known as the mouthbrooders, make no nest but carry the eggs, and

later the fry, in their mouths. This chore is usually undertaken by the female who picks up the eggs in her mouth as soon as they are laid, then sucks up the male's sperms to fertilize them. The male African cichlid *Haplochromis burtoni* spreads his anal fin to show off orange spots, the same size and colour as the eggs. The female mouths the fin, apparently trying to pick up these 'eggs', but gets a mouthful of sperms instead. The eggs develop in the parent's mouth where they are provided with a continuous current of water which she 'gargles'; this rubs the eggs together and keeps them clean.

The mouthbrooder family appears to avoid the perils that beset eggs, but this sophistication provides a niche for an unusual predator. Lake Tanganyika is one of several large African lakes with huge assemblies of cichlid species and in 1986 it was discovered that several are parasitized by the cuckoo catfish, which behaves like the familiar bird from which it takes its name. Eggs and fry of the catfish have been found only in mouthbrooding cichlids; none have been found in the open water. How the eggs get there is still a mystery, but it is presumed that they are picked up at the same time as the mouthbrooder's eggs. Like a cuckoo nestling, the newly hatched catfish demolishes its foster siblings, biting their bodies and sucking up the remains of the yolk or swallowing them whole.

For variety in egg-carrying techniques the amphibians are the specialists and they often use egg-

Far left: both male and female angelfish care for the family. The eggs are laid on a leaf after its surface has been picked clean. Both parents fan the eggs and tend the young fry when they hatch.

Left: once laid, the eggs are the responsibility of the male bullhead or miller's thumb. He will remain with them, to aerate them by fanning with his fins, until they have hatched. The eggs were laid on the underside of the stone which has been temporarily raised.

Top right: African mouthbrooders are fishes that protect their eggs and later their young by carrying them in their mouths.

Bottom right: the mouthbrooder *Haplochromis burtoni* spitting out a mouthful of babies after retrieving them as they strayed to the edge of the parent's territory.

carrying as a means of avoiding the necessity of returning to water to breed. The midwife toad has earned its name from its spawning habits. The male strokes the female to induce her to lay her eggs and, as she does so, she extends her hind legs to form a trough to receive them. The male fertilizes the eggs, then he plunges his legs into the 'trough'; as he does this the strings of spawn become wrapped around his legs and he carries them until they hatch. Other female frogs, such as the tree-dwelling Goeldi's frog of South America, have troughs or bowls on their backs to contain their eggs. The male Darwin's frog, also of South America, waits beside the eggs until they are near hatching. Movements of the tadpoles inside the eggs stimulate him to take them in his mouth and pass them into his vocal sac – the sac which fills with air and visibly swells when the frog is calling. There they stay through the entire tadpole stage until they are released as froglets.

All these egg-carrying frogs and toads lay fewer eggs than those species that lay their eggs in water and abandon them. The difference is considerable: egg carriers lay one to 50 eggs rather than the thousands of eggs laid by species that leave them to their fate. By putting their resources into parental care they achieve the same end as putting resources into laying large numbers of eggs. But perhaps the major benefit is the increased independence from water. Cutting out a trip to a pool for mating and spawning saves effort and reduces danger.

Who cares?

It is usual among mammals, including humans, for the female to assume the main burden of parental care. When the male wears the apron, so to speak, it comes as a surprise. A century ago it was regarded as almost improper. The giant waterbug carries its eggs on its back and it was originally thought that the carriers were the females. The record was put straight by a Victorian entomologist, a woman so thoroughly steeped in the prevailing prejudices concerning the roles of the sexes that she did not believe that a male could make a good nurse: 'That the male chafes under the burden is unmistakable; in fact, my suspicions as to the sex of the egg-carrier were first aroused by watching one in an aquarium which was trying to free itself from its load of eggs, an exhibition of a lack of maternal interest not expected in a female carrying her own eggs.' Prejudice had evidently coloured her interpretation of the bug's behaviour: without the male's care the eggs will not hatch. He swims to the surface at intervals so that the eggs can breathe and he frequently rubs the eggs with his hind legs, presumably to clean them.

When the mother supplies food for the newborn offspring from her own body, she is forced into a major parental role, but when she produces eggs containing all the nourishment needed for development, the physical connection is severed and it is then possible for either or both sexes to take responsibility for the young animals. Yet parental care remains the duty of the female in the great majority of cases. This can be explained by the unequal investment in the eggs. Females by definition produce the egg, the larger of the two gametes. So, although the sexes share the genetic investment in the next generation equally, the manufacture of an egg, even the tiny egg of a cod, requires a greater investment of energy than do millions of sperms. With internal fertilization, the female has no option but to continue her involvement with the egg: she cannot abandon it before it is laid. Consequently, she has good reason to continue to lavish care on the egg after it has been laid. In contrast, the male can improve his breeding success by abandoning a female after fertilizing her eggs and spending time and energy searching for other mates.

Examples of this unequal division of labour have already been given: female wolf spiders carry the egg cocoon and female solitary bees and wasps provision nests for their larvae. The male's role in reproduction ends in many cases with fertilization; fatally so in the case of the honeybee drone whose genital organs are ripped from his body when he tries to disengage from the female.

Nevertheless, there are many species in which the male plays an active and sometimes leading role in caring for the eggs. Sharing parental duties is the rule among birds, and the sight of common garden birds working together reinforces the view that this is the natural way to rear a family. Ninety per cent of bird species are monogamous. Apart from the monotreme mammals, they are the only group that lays very large eggs that need constant warming until they hatch. The strain of egg production and caring for the young is usually too much for the female to cope with alone and the male parent renders assistance by taking turns with incubation, by providing the female with food so that she need not leave the nest, or by guarding the nest against predators.

In some circumstances, for instance when the chicks feed themselves (as do pheasants and ducks), one bird is able to rear the brood on its own. The

Above: like all males of the pigeon family, the male Namaqua dove helps with parental duties. He brings material for the female to build into a nest, and he shares incubation of the eggs and feeding of the nestlings.

Right: a cock jungle fowl waits while one of his hens lays an egg in her well-hidden nest nearby. He actively helped her select the nest site, and while she lays he gives a special soft call. When she emerges, the hen may give the typical after-laying call and the cock will escort her so that she can concentrate on feeding while he keeps guard.

Bees invest in their offsprings' survival by building and provisioning a nest.
Top left: a leafcutter bee alights at the entrance of her burrow with a piece cut out of a rose leaf for building the cells in which she will lay her eggs.
Centre left: a cell removed from the burrow to show its construction.
Bottom left: the egg is laid on a mixture of nectar and pollen, which will nourish the growing larva.

remaining duties usually fall to the female but, in a few cases, the roles are reversed and she leaves the male to incubate the eggs. This can come about when an abundant supply of food means that the female can lay several clutches of eggs in quick succession. In parts of the Arctic where insects are plentiful, sanderlings lay two clutches and each parent incubates a clutch on its own. The spotted sandpiper, a North American species, can lay up to five clutches in a summer. When she lays only a single clutch she is monogamous and helps to incubate the eggs, but multiple clutches are incubated by the males. The female seeks out and courts each male in turn, lays in his nest and flies off – she shares the incubation only of her last clutch. The female northern jacana, found from Texas to Costa Rica, occupies a large territory in which several males build their nests and rear her clutches of eggs, while her contribution is to drive away predators.

Fewer than one per cent of bird species are polyandrous, but the habit is widespread among fishes. This raises an interesting question: why are female birds but male fishes the most assiduous parents? A survey of fish shows that the important factor is the method of fertilization. Parental care has been recorded in 54 of the 245 families of bony fish (excluding the cartilaginous fishes – the sharks, skates and rays). Of these, the male has sole responsibility for looking after the eggs in 30 families, in 28 of which external fertilization is the rule. In only two families with internal fertilization does the male show any concern for his offspring.

Where fertilization is internal, as in birds and mammals as well as some fish, the male can desert before the eggs are laid, and perhaps search for another mate. With external fertilization the situation often changes. As the female lays her eggs before the male fertilizes them, she can swim away while he is still engaged in this vital process. Sometimes, as in the Siamese fighting fish, the male has not only to fertilize the eggs, but also to catch each one in his mouth as it is laid and give it a coating of mucus so that it will float. Consequently, he is unable to depart until all the eggs are laid.

The male will be more closely tied to the eggs if he is guarding a territory which is also the site where the eggs will be deposited. But whatever the reason for the male becoming saddled with the eggs, once he has taken charge of them the scene is set for the male to evolve ways of enhancing their survival by guarding

and fanning them, removing dead eggs and building nests. This is precisely the situation with the three-spined stickleback, which is described on pages 73–4. The male has to defend a territory against other males if he is to attract females; having lured several to his nest in quick succession, he remains to care for the eggs and young. However, while the male stickleback goes to great lengths to care for his offspring, he must yield pride of place to the male sea horse and the related pipefish who not only carry the eggs and young in special pouches like kangaroos but also nourish them.

During courtship the male sea horse 'bows', bending his body to expel water from the pouch on his belly. The female inserts her ovipositor into the opening and lays her eggs which become embedded in a spongy lining, provided with abundant blood vessels, that develops prior to courtship. The lining acts like the placenta of a mammal; it exchanges oxygen and carbon dioxide and, in the pipefish and probably the sea horse, it also secretes a nourishing fluid that supplements the yolk. These fishes have almost completely reversed the usual role of the sexes and demonstrate that 'motherhood' is not inevitably the role of the female.

Incubation

Embryonic development is, like all other biological processes, dependent on temperature: as temperature rises, metabolism speeds up. Consequently, keeping the eggs warm is an advantage because development time will be shortened, the chances of the eggs being eaten by predators are reduced and the strains of guarding them minimized. Eggs may be warmed passively, as in the case of frogspawn which warms up in the sun, but the term incubation implies active steps taken by the adult, as with the pirate spider carrying its eggs into a sunny position and the American alligator making a nest of rotting vegetation that warms up like a compost heap to heat the eggs.

Incubation performed by the parent sitting on its eggs to warm them with its body heat has evolved several times. It is an important feature of the life of birds: everyone knows that a bird's egg will fail to hatch if it becomes chilled. Some other animals, while not warm-blooded in the sense of generating body heat by metabolism, are nevertheless able to raise their body temperatures and incubate their eggs. Bees could be said to incubate their eggs after a fashion because the hive is kept warm by the workers.

Bumblebee eggs die if the temperature of the nest drops below 10 degrees centigrade and they are normally maintained between 30 and 32 degrees. Until the first batch of workers grows up and takes over the duties of foraging and warming the nest, a queen bumblebee has to incubate the eggs with heat transferred from her almost hairless underside; it is important that she does not spend too long away from the nest, especially in cold weather, lest they cool too much.

Bees can incubate their eggs because, like many large insects, they have to warm their muscles to around 30 degrees centigrade to make flight possible. Some reptiles are capable of the same feat; they raise their body temperature either by basking in the sun or by producing heat by metabolism. The pythons are known to use a high body temperature for incubation. The female snake coils tightly around her clutch and shivers to raise her temperature to 32 or 33 degrees centigrade. Pythons are a largely tropical group, but the diamond python extends from tropical Australia southwards to the latitude of temperate Canberra and also inhabits cool mountain slopes, presumably because of this ability to incubate its eggs.

In recent years the temperature of reptile nests has been found to have an unusual significance. Mississippi alligators living in the swamps of Louisiana lay their eggs in a variety of places. Eggs from nests on dry embankments almost always hatch into males, whereas those laid in wet marshy ground usually hatch into females. The controlling factor is temperature. An alligator egg has the genetic potential to develop into either sex, but those incubated above 34 degrees will develop into males, while those incubated below 32 degrees develop into females.

A similar mechanism of 'environmental sex determination' has been found to operate in lizards, tortoises and turtles, as well as in various fishes. The painted and map turtles of the Mississippi lay their eggs in the sand. If the nests are in the shade of bushes, the eggs develop into males, while those that are exposed to the sun's glare produce mostly females. There is no evidence that females choose the sex of their offspring by selecting an appropriate nesting site; neither is the advantage of this form of sex determination apparent. However, environmental sex determination in reptiles has important implications for the conservation of endangered species. The survival of the young is improved by

removing the eggs from vulnerable nests, incubating them in a safe place and releasing the hatchlings. Unless the critical temperature for deciding sex is known and batches of eggs incubated accordingly, a distorted sex ratio may be produced.

With the evolution of true warm-bloodedness in the birds, incubation became a necessity. In a warm-blooded animal the embryo is as dependent on a high temperature as its parents: development ceases below 25 to 27 degrees, and the embryo is unlikely to hatch out if incubated below about 35 degrees. Continuous transfer of heat to maintain a temperature of only a few degrees less than the adult blood temperature is essential throughout the development period. This is not only expensive in terms of the parents' time and energy but needs carefully co-ordinated behaviour on the part of the parents.

With few exceptions birds transfer their body heat to the eggs by means of a brood patch. This is an area of skin on the underside of the body where the feathers are shed before egg-laying and a rich supply of blood vessels acts as a hot-water bottle. A few birds lack brood patches: the gannets wrap their webbed feet around their eggs and the swans, ducks and geese pluck feathers from their breasts to line their nests.

Above: a mallard duckling about to emerge from its egg. Several days earlier, it started to listen to its mother's voice and learnt to recognize her, so it will stay close to her when she eventually leads it from the nest.

Far left: loud chirruping from inside her eggs causes the Nile crocodile to uncover the nest and gently crush the eggshells to help the baby crocodiles hatch out. Then she picks the tiny hatchlings up in her huge mouth and carries them to the water.

Left: the flying gecko of Malaysia lays her round white eggs in pairs, cementing them under the bark where she hides during the day. Gecko eggs have brittle shells like birds' eggs.

Right: the female greylag goose has sole responsibility for incubation, but her mate stays on guard nearby and both parents lead the goslings from the nest.

Below: a dunlin sits tight on its nest, despite approaching danger. It relies on camouflage to avoid detection and flies away only at the last moment.

An incubating bird is a miracle of patience. It may have to sit tight for many days, in the case of albatrosses and related sea birds; it may also be plagued by flies and unable to lift a foot to scratch itself. It may have to stay quite still as a predator approaches, to keep the location of its eggs secret. But every now and then it appears to be restless. It gets up, peers down at its eggs and pokes among them with its beak. It is almost as if it is trying to get comfortable, but in fact it is checking the wellbeing of the eggs. To regulate the temperature of a clutch of eggs within a fine limit of a few degrees requires that the sitting bird does more than simply sit on the eggs to cover them. It must monitor the egg temperature and make allowances for changes in the environment and also for the heat produced within the eggs. The metabolic processes taking place in the embryo generate heat which will increase as it grows and so reduce the heat needed from the parent. Towards the end of incubation the embryo in a gull's egg is producing one fifth of the heat it needs.

Birds with large clutches, tits in their nest holes or mallards among the rushes, cannot warm all their eggs effectively; a difference of several degrees can develop between the eggs in the centre and those at the edge of the nest, so the sitting bird shifts them, poking and rolling them with its beak to rearrange them. The movements have a second and equally important function. It has long been known by poultry breeders that eggs in an incubator must be turned at intervals if they are to hatch. In the early part of development, turning prevents the embryonic membranes from fusing with the eggshell membranes. If this happens the embryo will stick to the shell and development can be fatally distorted or the chick may find itself in a position that prevents it from hatching out.

At a later stage, the two sets of membranes do fuse and the position of the embryo in the egg becomes fixed. It must now stay the right way up for hatching and this is achieved by the embryo becoming 'bottom heavy', which means that the egg will naturally come to rest in a particular position. When the bird pokes among the clutch, it is separating each egg from

Left: a blue tit has difficulty in keeping her large clutch warm, but she is helped by the cosy nest lined with feathers.

Below: a feral pigeon shifts its eggs in the nest. This action is carried out at intervals to ensure that the eggs are warmed evenly and will develop properly.

contact with its neighbours so that it can roll into the correct orientation. Just before hatching the chicks in the eggs call loudly if they are turned upside down and their cheeping stimulates the parent to shuffle the eggs. Yet despite all the evidence that egg-turning is necessary, a few birds manage without it. The palm swift, for example, glues its egg to a palm frond, making egg-turning impossible.

To reduce the strain of incubation, energy can be saved by insulating the nest through adding extra material or building it in a cosy hole or crevice. This will allow the parents more time to feed, or to bathe and preen. Additionally, it enables the birds to extend their breeding ranges. The brambling is a finch very similar in many respects to the closely related chaffinch, but it extends further into northern Europe with the help of a better insulated nest, and hummingbirds nesting on the slopes of the Andes where the air is cool make deep, well-lined nests compared with those in the lowland forests below.

The importance of a correct incubation regime was neatly demonstrated by the introduction of two bird species into North America. Starlings were brought to New York from temperate Europe in 1890 and had spread across the continent by 1960. However, several crested mynahs from tropical Asia escaped in Vancouver in 1895, but the species has still not spread. Both birds nest in holes, but the mynah has the tropical bird's habit of leaving its nest in the heat of the day – a regime that kills most of its eggs in temperate America. As an experiment, some mynahs were given heated nestboxes and their hatching rate improved.

While tropical birds can leave their eggs uncovered for some time because they will cool only slowly, they may also face the possibility of dangerous overheating. Nests with roofs or those built in holes provide shelter from the sun, but birds with exposed nests stand over the eggs to shade them. In the hottest conditions, where air temperature rises above body temperature, desert birds revert to sitting on their eggs, but they now act as a cold compress rather than a hot-water bottle, drawing heat away from the clutch.

The Egyptian plover uses another ploy: it shelters its eggs by burying them in the sand, which it wets with water brought in by soaking its breast feathers in a nearby river.

Many birds share incubation, with the male and female taking alternate shifts. These vary in length from less than an hour in small birds to days among sea birds which have to travel long distances from the nest in search of food: wandering albatrosses nesting on the South Atlantic island of South Georgia have incubation shifts of five to ten days and, when off duty, they travel hundreds of kilometres to find food.

The emperor penguin, which nests on the coasts of the Antarctic continent, has a peculiarly difficult problem. It is the largest of the penguins and, to rear its single chick to independence before the winter sets in and the seas freeze, it has to lay its egg in the middle of the previous winter. As soon as she has laid, the female emperor penguin passes the egg to her mate and walks away in search of open water where she can dive for squid. The male, who has no nest, balances the egg on his feet and covers it with a fold of skin to keep out the intense cold. Then he shuffles over the ice with the other males, trying to find shelter from the worst of the Antarctic winter weather. The female is absent for the whole of the nine-week incubation period. It is a long walk, as much as 100 kilometres, across the frozen sea; when she reaches open water she has to eat enough to make up her lost weight, as well as find food to bring back in her crop to the chick which will be ready to hatch on her return.

The incubation regime of the emperor penguin suits its unique environment. The more common practice is for the female of the species to take over the whole of incubation and this is reflected in her often drab plumage, suitable for camouflage, compared with the male's bright courtship dress. Single-parent incubation places a greater strain on a bird because it is difficult to feed properly. Many birds lose weight during incubation and it has been known for snow geese to starve to death on the nest. The situation can be alleviated by the male bringing food to his mate, in an extension of the courtship feeding that had earlier helped her form the eggs within her body. Females that are fed include the finches and tits, which consequently need to leave the nest to feed less frequently, and birds of prey, which do not leave the nest at all. The extreme of provisioning by the male is shown by the hornbills that wall up the female in the nest hole and feed her through a small opening.

There are several ways of making a significant reduction in the energy cost of incubation. The cuckoos and other brood parasites get other birds to incubate their eggs, and a few birds use heat from the environment. The grebes, which have rather low body temperatures, are helped by the heat generated by their nests of rotting vegetation. One group of birds, the megapodes (meaning 'big feet'), also known as the incubator birds or mound-builders, have abandoned conventional incubation entirely and are worth considering in detail because their nesting

Far left: under the midday African sun the body temperature of the crowned plover is lower than that of its surroundings, so the parent covers its eggs to cool them rather than to warm them.

Left: while one brown skua sits on the nest, the other stands guard and will attack humans or other large animals that come too close.

Right: when incubation is shared, it is vital for the parents to co-ordinate their behaviour. For example, a wandering albatross cannot leave its nest until its mate has returned to relieve it. Displays, such as this 'sky-pointing', help to promote the bond between the pair.

behaviour is so different from other birds.

Megapodes live in Australia, New Guinea, Indonesia and Polynesia. Like the turtles, the maleo fowl lays its eggs in sandy beaches, burying them at a depth where the temperature is consistently high. The scrubfowl employs volcanic heat at hot springs by digging a 1–2 metre pit to find soil warmed to around 37 degrees centigrade. Where there are no sunny beaches or volcanoes, the megapodes make use of rotting vegetation as a source of heat. Some scrubfowls simply lay their eggs in rotting logs, but other species scrape up huge mounds of rotting leaves. The brush turkey's mound consists of more than $3\frac{1}{2}$ metric tons of leaves and soil, a bulk that makes the internal temperature very stable. The mallee fowl goes to enormous lengths to keep the nest temperature steady. Throughout the period of incubation the male is in attendance, ensuring that the temperature of the nest cavity in the centre of the mound is at an even 33 degrees. He tests the temperature by taking soil samples in his beak and, accordingly, opens the mound when rotting becomes too fierce, piles on extra soil when the sun is too hot or scrapes away soil to expose the eggs to a weak sun.

The strange behaviour of the megapodes is not fully understood. Why, for example, have they reverted to the habits of their reptilian ancestors? The answer may be that this is one of the birds' many attempts to save energy. Using environmental heat is a neat solution. Some megapodes can abandon their eggs as the chicks are so precocious that no parental care is

required, but various adjustments have had to be made in the physiology of the egg. It has to develop in an atmosphere that would kill other birds' eggs. The nest pits are exceptionally close and stuffy: the atmosphere is virtually saturated with water vapour, while oxygen concentration is low and carbon dioxide is at a level that would also be fatal to other eggs. This remains a problem for the physiologists. Another conundrum is that the egg-turning that is so vital for most birds is impossible. The eggs are positioned vertically in the nest and presumably the embryo automatically assumes the correct position for hatching. The high humidity of the nest prevents water loss and the formation of the airspace. This means that megapode chicks cannot pip in the same way as other chicks and they can only start to breathe when the beak is thrust through the shell.

The elaborate behaviour of the mallee fowl contradicts the energy-saving theory. The eggs are laid over a period of four months; each takes two months to hatch and with the time spent constructing the nest the male is kept busy for 11 months of the year. The benefit is that the mallee fowl has been able to spread from the evenly hot tropical climate enjoyed by the rest of the megapode family into the cooler regions of Australia. However, the megapodes have also discovered the ultimate energy-saving parental strategy of brood parasitism, which involves getting something for virtually nothing: some scrubfowl lay their eggs, cuckoo-like, in the mounds of other megapodes.

CHAPTER 10

Beyond the egg

In the *Chronicles* of the Tudor writer Raphael Holinshed there is an account of how, in times of danger, the female adder swallows her offspring. This seems to have been a long-standing legend and its origin is not difficult to guess: someone cut open a female adder and found unborn young inside. The adder is, like many snakes and lizards, ovoviviparous – the eggs are retained in the female's reproductive tract until they are fully developed and hatch at the moment of laying. This is little different from carrying the clutch after laying. The eggs are protected from desiccation and egg-eaters, and, as will be discussed later, the female can speed their development.

There is no real distinction in the mechanics of development between ovoviviparity and egg-laying or oviparity, except in the ability to retain the eggs in the oviduct. It is not unusual for reptile eggs to commence development before laying and the ovoviviparous species have merely developed the tendency, thereby sparing themselves the necessity of giving the eggs a full shell.

Retaining the eggs in the body does, however, set the scene for a major advance in reproduction: true live birth after a period of gestation during which the embryo is nourished from the maternal bloodstream instead of being provisioned with yolk before development starts. This is known as viviparity and one of its advantages is that the female no longer has to find enough food to supply the clutch at one instant; the burden of nourishing the embryos can be spread. She can also increase the supply of food without having to lay impossibly large eggs and enable the embryos to continue development and growth beyond the stage that can be reached with a fixed supply of yolk.

Live birth has evolved a number of times across the animal kingdom, but it has become a distinguishing feature only of the marsupial and placental mammals. Its advantages are offset by some drawbacks which, in the case of the birds, have proved overwhelming: no bird has made any move towards giving birth to live young. The pros and cons of live birth can be

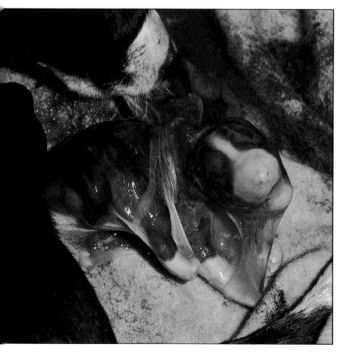

illustrated best by comparing the mammals and birds and tracing their evolution from the reptiles among which several experiments in live birth have been made. But before embarking on this, a survey of the animal kingdom shows that live birth is not the prerogative of the 'higher animals' and has been attempted by a variety of unlikely species. Several have evolved some form of placenta, remarkably like that of the mammals, to link the maternal and embryonic blood systems; others have found their own unique techniques.

In view of the many and varied adaptations that the insects have evolved to take up an amazing selection of lifestyles, it is not surprising to find that viviparity has evolved several times in this group. There are two basic systems of gestation: either the egg is retained and nourished in the reproductive tract until it hatches, or the egg hatches and the larva is retained and nourished. In both systems there is a connection between the maternal and embryonic tissues.

The most familiar live-bearing insects are the aphid pests of farm and garden. They achieve a phenomenal rate of reproduction not only by parthenogenesis, which eliminates the need to find a mate (see pages 44–6), but also by telescoping the developmental period. The eggs have no shells and are retained in the ovary where they absorb the necessary nutrients through its lining. The embryos develop rapidly and are born as nymphs – immature insects resembling the adult – which accumulate around the mother as a small flock of young aphids.

Giving birth seems very different from laying eggs, but every mammal comes from an egg cell. The essential difference between a live-bearing mammal and an egg-laying bird lies in the way the embryo is nourished.

Far left: a dog is clearing the membranes from a newborn puppy and (*left*) feeds her puppies on milk.

Above right: the water flea *Daphnia* with a brood of young inside its shell or carapace. Under certain conditions the eggs are retained until hatching and the embryos feed on secretions from the mother's body.

Right: an aphid gives birth to live young. The eggs are retained in the ovary where they receive the necessary nutrients for their development.

145

The tsetse fly is another devastatingly effective live-bearer. Despite a slow breeding rate of a dozen young in a six-month lifetime, it prevents the development of huge areas of Africa by spreading sleeping sickness among humans and a disease called nagana among cattle. The live-birth system differs from that of the aphid in that the normal-sized, yolky egg hatches inside the insect and the larva is retained until it is ready to pupate. Only one larva is carried at a time and it grows until it takes up most of the space inside its mother's body. During the gestation period it takes nourishment from a gland that absorbs nutrients from the maternal body and secretes them through a 'nipple' in the oviduct wall, close to the larva's mouth. Respiration is achieved by the larva extruding the breathing tubes on the tip of its abdomen through its mother's genital opening.

Among the insects' relatives, scorpions habitually give birth to young which climb onto the mother's back until they can fend for themselves. Some scorpions are ovoviviparous, retaining a yolky egg until it hatches, but others are viviparous, producing an almost yolkless egg. As the embryo of a viviparous scorpion develops it becomes attached to the maternal intestine by an 'umbilical cord'. More remarkable is *Peripatus*. This soft-bodied, caterpillar-like creature, which lives in damp places mainly in the southern hemisphere, was until recently thought to be a 'missing link' between the arthropods (including the insects, spiders and crustaceans) with their jointed legs, and the annelid worms (such as the earthworms, lugworms and ragworms). It is now realized that *Peripatus* has no close link with the worms, but it is still interesting as an arthropod that has primitive characteristics which are sometimes combined with an advanced system of live birth. Some species lay normal, yolky, shell-covered eggs, but others retain eggs that absorb nourishment from the oviduct lining, while a third group has small eggs that develop into embryos, each of which is surrounded by an amniotic membrane and attached to the oviduct via an umbilical cord and a placenta with similar functions to those in mammals.

These examples of live-bearing invertebrates have come from animal groups which practise internal fertilization and lay their eggs on land. Among vertebrates, the ability to give birth to live young appears only fleetingly among the fishes which are fully aquatic, yet there is the same gradation from oviparity to viviparity as in the insects, with normally oviparous species sometimes holding back egg-laying until development has started or retaining their eggs until development is complete and finally progressing to the point where the embryos are nourished internally.

The full range is illustrated by the family of topminnows. These are small freshwater fishes, also known as toothcarps, which are familiar to aquarists; the best-known species is the guppy. The egg-layers comprise the killifishes, which include the annual fishes that complete their life cycle within one rainy season and survive to the next as eggs, and the rice fish of Japan that retains its eggs while development starts but is still oviparous.

Left: a tsetse fly gives birth to an enormous, fully developed larva. The egg had hatched inside the fly's reproductive organs and the larva was retained and nourished until it was ready to pupate.

Above: a scorpion with its brood of young. Some species are ovoviviparous, retaining the eggs until they hatch; others are viviparous or live-bearing and nourish the developing embryo internally.

Left: *Peripatus*, sometimes called the velvet worm, is oviparous, ovoviviparous or viviparous according to species. This Jamaican species is giving birth to live young which it has nourished by means of a placenta that is remarkably similar to those of mammals.

147

The mollies, platys and swordtails and the guppy are ovoviviparous or viviparous topminnows. In these species the male is equipped with an intromittent organ, or gonopodium, fashioned from the anal fin, and fertilization is internal. The guppy is ovoviviparous and the eggs develop while still in the ovary. The live-bearers, such as the dwarf topminnow and the four-eyed fish, also keep their eggs in the ovary but they have a placenta-like system, differing in details between species, for transferring nutrients from the mother to the embryos.

Live-bearing is rare among the bony fishes, which make up the vast majority of the world's fishes, but it is common among the cartilaginous fishes – the sharks, skates and rays – where there is the now familiar progression from oviparity to viviparity. Egg-laying species deposit each egg in a horny cocoon which may later be washed up on the beach as a 'mermaid's purse'; ovoviviparous rays supplement the yolk with a milky secretion which the embryo takes in through its mouth or absorbs through its gills, and the live-bearing butterfly rays have long tubes in their oviducts which enter the embryos' spiracles, or 'nostrils', and pass nutrients into the gut.

The blue shark, hammerhead shark and some dogfishes have a placenta formed from the yolk sac which fuses with the oviduct lining in a remarkable parallel with the mammalian placenta. At the point of attachment, the egg membrane disappears and the outer tissue of the yolk sac breaks down so that blood vessels of embryo and mother come into close contact.

Among amphibians, a group of animals taking the first steps towards a terrestrial lifestyle, internal fertilization is rare so moves towards live-bearing are not to be expected. Internal fertilization is, however, practised by the newts and salamanders, and the black salamander of the Alps mates on the snow in early spring then gives birth to live young without entering water. The tadpoles develop within the mother and the gills act as a placenta. In contrast, the Surinam toad retains an aquatic lifestyle, complete with external fertilization, but it has a unique method of external gestation. The female carries her eggs in pockets on her back until they hatch out as tadpoles or froglets, depending on the species. To get the eggs in place, the male grasps her around her middle and the pair loop-the-loop in the water. As they go over the top of the loop, upside down, the female sheds a few eggs and the male catches them on his belly. Then, as the pair come out of the bottom of the loop and the male is on top again, the eggs are pressed against the female's back where they stick in place. The operation is repeated until all the eggs are laid, then the pair separate. Over the next few hours, the female's back swells and encloses the eggs in pockets capped with lids. Secure within its pocket each embryo develops into a tadpole or froglet, using its tail as a placenta for absorbing food from the mother during the tadpole stage.

The differences between egg-laying and live-bearing become crystallized in the amniotes, the three highest groups of animals. The reptiles are essentially egg-layers, but many species (such as the adders

Swordtails are live-bearers that are popular as aquarium fish.
Far left: a female is giving birth to a baby whose head can be seen emerging.
Left: as the baby comes free, the mother jerks away and it swims to the surface. Here it swallows air to fill its swimbladder, which enables it to float easily.

Right: a marsupial frog, *Gastrotheca ovifera*, gives birth to a brood of 37 froglets. The eggs are deposited in a pouch on her back where the tadpoles remain, with the gills acting as a kind of placenta, until they have metamorphosed.

Below: the pygmy marsupial frog *Flectonotus pygmaeus* releases its young as tadpoles; they complete their development in the tiny rainwater pool formed by a bromeliad leaf.

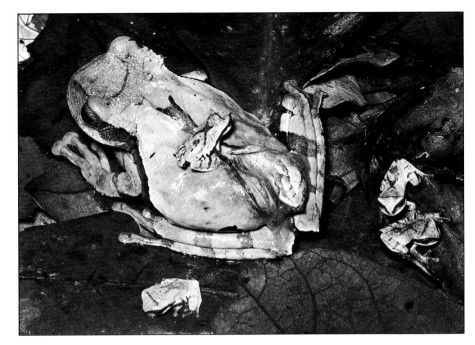

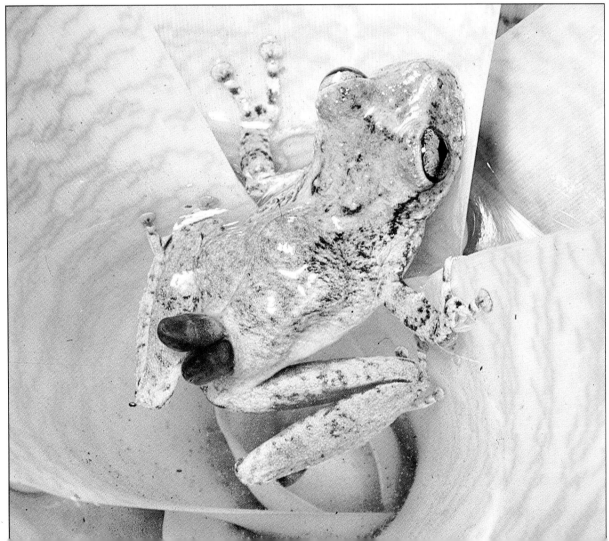

mentioned at the beginning of this chapter) have taken to retaining their eggs for at least part of their development and full live-bearing has evolved more than 30 times in different families of lizards and snakes. The birds and mammals have evolved separately from the reptiles and the birds have remained confirmed egg-layers while nearly all mammals have abandoned egg-laying in favour of live-bearing.

Among reptiles, viviparity is a feature of species living in cold climates, possibly because it is a means of speeding development through the female basking in the sun to raise her body temperature. As a first condition of viviparity the hard eggshell is lost, enabling the embryonic membranes to make contact with the maternal tissues. This occurs in the common lizard, but only water and dissolved gases are transferred; the embryo still relies on yolk for its nourishment. A more intimate connection between mother and embryo allows food material to be transferred. For example, in the European skink (a

kind of lizard) and the Australian copperhead snake the oviduct and the embryonic membranes are thrown into folds to increase the area of contact between them; the tissues are so thin that where they meet dissolved substances can pass easily in both directions.

The mammals have pursued the theme of viviparity. Stripped of fur and scales, a basic mammal does not look so different from a basic reptile – many species in both groups are simply four-legged, long-tailed animals designed for running on the ground – but there have been major changes in reproduction. Live-bearing and lactation are two of the main features of mammals, whose name was coined by the Swedish naturalist Linnaeus from the Latin word *mamma* – a breast. The first mammals were oviparous and their only surviving representatives, the platypus and echidnas or spiny anteaters, lay and incubate large yolky eggs with leathery shells. All other mammals are live-bearers which produce small, almost yolkless eggs that make contact with the

Left: the slow-worm is an ovoviviparous, legless lizard whose eggs hatch as they are laid.

Right: live birth is commonly a feature of reptiles living in cool climates. These viviparous African chameleons were born in the tropics, but at a high altitude where the nights are cold.

maternal tissues during cleavage. The details of mammalian cleavage differ from the process described for amphibians and birds in Chapter 7 because of this need to create a placenta and receive nourishment from the mother as soon as possible. The placenta is fashioned, in different mammal groups, from the membranes of the yolk sac, the chorio-allantois or both. After giving birth, the female mammal continues to be tied physiologically to her offspring by lactation, assuring them of a steady supply of food until they are capable of feeding themselves.

The birds, it has been suggested, are in reality living dinosaurs, albeit somewhat different from the popular conception of these animals. In their evolution from the reptiles, they have undergone a radical restructuring of their bodies in order to become efficient flying machines, yet their method of reproduction remains extremely conservative and there is not even a tendency to retain the eggs until development is well advanced. The explanation for this conservatism, which sets the birds apart from the other major groups of animals, must be related, along with their other characteristics, to the demands of flight. However, even the ratites (the ostrich, emu, kiwi, cassowary and rhea), which gave up flight early in their evolution, have made no move towards viviparity.

Egg-laying saves weight by removing the necessity of carrying not only the developing embryo and its placenta but also an enlarged reproductive system to house them. Yet even the relatively light weight of an egg passing down the oviduct can be a burden to some female birds which may be reluctant to fly immediately before laying. (The bats, which compete for dominance of the air, are live-bearing, but no bat gives birth to more than one baby at a time.) The birds' main advances in the breeding process have been in the period after hatching where, compared with reptiles and all other animals except mammals and social insects, parental care has assumed a great importance.

Despite the enormous difference between, for example, a hen laying eggs and a cat giving birth to kittens, the essential processes of reproduction, of transforming a fertilized egg into an animal, are not vastly different. The difference lies mainly in our subjective view of the hard-shelled egg, beautiful in its stark symmetry but outwardly inanimate, contrasting with the animated, mewling kitten crawling towards its mother's teats. The gulf between them seems so wide that the scientific community of the nineteenth century was unwilling to accept that a mammal could lay eggs, even so strange a mammal as the duck-billed platypus recently brought back as specimens from Australia. The naturalist W.H. Caldwell eventually sailed to the 'new' continent to settle the matter, and in 1884 he electrified a meeting of the British Association for the Advancement of Science with the laconic telegram: MONOTREMES OVIPARUS OVUM MEROBLASTIC (monotremes egg-laying, egg with partial cleavage). Caldwell had found a female platypus on the point of egg-laying and discovered that the embryo develops as a disc of cells lying on top of a mass of yolk, just as in birds and reptiles.

To sum up, the eggs of birds and mammals, or of any other animal, show little difference in the essentials of their development. Nature has little aptitude for innovation and chose the egg as the device for the renewal of life and for introducing variety by means of sex. Once a workable method of turning the egg into an animal had been achieved, there was no need to change it. The enormous outward difference in the eggs of a mammal, a bird or a sea urchin is the expression of the way that the embryos are supplied with food. The sea urchin egg is launched with little investment from the parent. It has a small supply of yolk and quickly becomes a larva that feeds itself while development is completed. The mammalian egg is also small, but it is sheltered in its mother's body and is quickly provided with food from outside. The bird egg, by contrast, is provided with plenty of yolk to nourish it through development and the embryo remains inside until it reaches a stage of development comparable to that of the newborn mammal.

All eggs fit into one of these three patterns. Thereafter, the immense outward variety of eggs paraded on these pages is an expression of the need to ensure their survival and to promote the survival of the animals that will hatch out of them. Innovation may be scarce in nature but, once a theme has been established, the possibilities for expression of variation in the form and care of eggs is as boundless as the diversity of the millions of species that produce them.

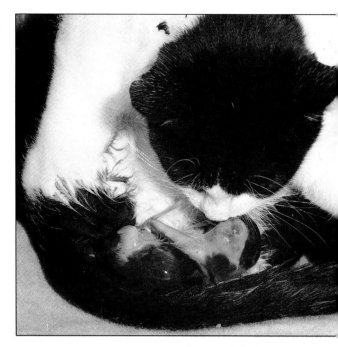

Embryonic development is similar in live birth and egg-laying but, compared with the laying of a clutch of birds' eggs, the birth of a litter of baby mammals is a very involved process. *Above left*: a kitten is born but the placenta has still to be delivered. *Above centre*: the umbilical cord and the placenta are the key to the difference between live-bearing and egg-laying. *Above right*: live birth in mammals involves greater participation by the mother. This cat is licking her kitten to remove the remains of the amnion. *Above*: she then eats the placenta to recoup some of the nutrients she has lost in pregnancy. *Right*: having nourished the embryos internally, through the placenta, the mother continues to feed her offspring after their birth.

Photographers' postscript

The photographs in this book represent three years of dedication to eggs – not only to the eggs themselves, but to the animals that produced them and to those that hatched from them.

There is a fascination about eggs. Their austere geometric shape when newly laid belies the complexity of the embryo that will develop within. Many times we watched as the egg contents turned into a living animal that was not only perfect in form but also knew exactly how to behave. Watching this process closely from day to day in numerous different types of egg impressed on us its miraculous nature. No marvel of man's technology can ever begin to approach what is contained in a living egg.

To photograph the development and hatching of eggs we tried to use species that are regularly bred in captivity, since the disturbance of eggs and nests in the wild is not only undesirable but unlawful. Much of our time was therefore taken up with maintaining animals in and around our home and in caring for the eggs they produced. A small incubator coped with birds' eggs that were 'borrowed' from tolerant mothers, such as a Muscovy duck and a jungle fowl hen, so that their hatching could be photographed. The hatchlings were then returned to the nest to resume their normal lives. A white garden fantail pigeon 'loaned' us one of her two new-laid eggs which we incubated and 'candled' daily to show development. After the chick had also been photographed through the stages of hatching, it rejoined its sibling in the nest to be reared by its parents. Other incubator-hatched birds were reared by broody bantam hens or in an artificial brooder. At various times our airing cupboard was an artificial brooder and rearing pen for many baby birds, as well as a nursery for the baby hedgehog which later ate (in front of the camera) an addled pheasant's egg. It was also a hatchery for reptile eggs (such as the large clutch of grass-snake eggs revealed when the compost heap was turned); and it was a breeding ground for generations of cockroaches, brine shrimps and tsetse flies.

Few invertebrates are regularly bred in captivity, so we had to search for their eggs in the surrounding countryside and photograph them *in situ* or bring them home to hatch under controlled conditions. Removing eggs from the wild requires care and understanding. The eggs must be kept under optimum conditions and the hatchlings returned to suitable habitats afterwards.

Caring for eggs is a responsibility and a challenge and we endeavoured to ensure that all the eggs we photographed hatched successfully and that the young were given a good start in life. The only exception was the hen's egg, since it was necessary to break open the shell to reveal details of the embryo. It seemed initially that destruction of a hen's egg was inconsequential, considering the millions that are daily cracked into frying pans, but opening one to find a living embryo inside evokes a feeling of remorse, the intensity of which is in direct relation to the stage of development of the embryo.

All baby animals are extremely vulnerable at the time of hatching. If conditions are not exactly right, if they get too warm, too dry, too damp, too cool, they may be unable to hatch out and will die. Baby birds are especially vulnerable to drying and chilling and may end up dead-in-shell or emerge as cripples if not adequately protected. We are pleased to be able to record that none of the other baby birds died as a result of our photography; all were reared safely and many are with us still.

Further reading and acknowledgements

Further reading

Alexander, R. MacNeill,
The Chordates, Cambridge University Press, 1975

Austin, C.R. and Short, R.V.,
The Evolution of Reproduction, Cambridge University Press, 1976

Bone, Q. and Marshall, N.B.,
The Biology of Fishes, Blackie, 1982

Calow, P.,
Invertebrate Biology, Croom Helm, 1981

Campbell, B. and Lack. E., eds,
A Dictionary of Birds, T. & A.D. Poyser, 1985

Grant, Philip,
A Biology of Developing Systems, Holt, Rinehart and Winston, 1978

Hinton, H.E.,
Biology of Insect Eggs, 3 Vols, Pergamon, 1981

Hughes, R.N.,
A Functional Biology of Marine Gastropods, Croom Helm, 1986

Oppenheimer, S.B.,
Introduction to Embryonic Development, Ally and Bacon, 1980

Romer, A.S.,
A Shorter Version of the Vertebrate Body, W.B. Saunders Company, 1956

Sadleir, Richard M.F.S.,
The Reproduction of Vertebrates, Academic Press, 1973

Slack, J.M.W.,
From Egg to Embryo, Cambridge University Press, 1983

Waddington, C.H.,
Principles of Embryology, George Allen and Unwin, 1956

Yapp, W.B.,
Borradaile's Manual of Elementary Zoology, Oxford Medical Publications, 1955

Author's acknowledgements

Information on the natural history and biology of eggs was gathered from the books listed here and from numerous scientific journals; I am indebted to their many authors. I am also grateful to the following for assistance: Richard Burton, Andrew Clarke, Colin Herbert, Sinclair Lough and Ian Swingland.

Photographer's acknowledgements

Jane Burton and Kim Taylor would like to acknowledge their thanks to:

Busbridge Lakes Ornamental Waterfowl (for point-of-hatch eggs); Charles Botting (for trout eggs); Kathryn Baker (for tortoise hatch); Ray White (for partridge nest); and Dr A.M. Jordan and Malcolm Flood, Tsetse Research Laboratory (for tsetse flies).

Picture credits

Illustrator: Rodney Paull
All photographs by Jane Burton and Kim Taylor except as follows:
12 Peter Parks, Oxford Scientific Films; **15** Peter Parks, Oxford Scientific Films; **25B** Anthony Bannister, NHPA; **31BR & 33** Michael Fogden; **35** G.I. Bernard, Oxford Scientific Films; **75BR** Robert Burton; **80–81** Dr. J.A.L. Cooke, Oxford Scientific Films; **89TL, TR, BR** Robert Burton; **93TR, B** Michael Fogden; **125** Michael Fogden; **138** Jonathan Blair, Susan Griggs Agency; **140B** Robert Burton; **142R** Robert Burton; **143** Robert Burton; **147T** N.M. Collins, Oxford Scientific Films; **147B** Dr. J.A.L. Cooke, Oxford Scientific Films; **149** Michael Fogden.

Index

Page references in italics refer to illustrations.